Linear Algebra Based Controllers

Gustavo Scaglia • Mario Emanuel Serrano
Pedro Albertos

Linear Algebra Based Controllers

Design and Applications

Gustavo Scaglia
Instituto de Ingeniería Química –
Departamento de Ingeniería Química
Consejo Nacional de Investigaciones
Científicas y Técnicas (CONICET),
Universidad Nacional de San Juan
San Juan, Argentina

Mario Emanuel Serrano
Instituto de Ingeniería Química –
Departamento de Física
Consejo Nacional de Investigaciones
Científicas y Técnicas (CONICET),
Universidad Nacional de San Juan
San Juan, Argentina

Pedro Albertos
Depto. Ingeniería de Sistemas y
Automática – Instituto Universitario de
Automática e Informática Industrial
Universitat Politècnica de València
Valencia, Spain

ISBN 978-3-030-42820-4 ISBN 978-3-030-42818-1 (eBook)
https://doi.org/10.1007/978-3-030-42818-1

This Springer imprint is published by the registered company Springer Nature Switzerland AG
The registered company address is: Gewerbestrasse 11, 6330 Cham, Switzerland

To God, Lourdes, Amanda, Juan, and Ema.
To God, Laura, Santiago, Joaquín, and Josefina.
To Ester, Leticia, Eduardo, and David.

Foreword

It is a great honor for me to write this foreword for a book I was expecting to find as a reference. I am very grateful to Dr. Scaglia, Dr. Serrano, and Dr. Albertos for their generosity in sharing their knowledge.

A book can be appreciated for many reasons. One may be its valuable content and the knowledge it provides. Another reason could be because it is the extract of the experience of people who have worked a long time on the subject and have made valuable contributions in this field. Both apply in this case. My experience in this field started, in some way, suddenly with my doctoral thesis, guided by Dr. Scaglia and Dr. Serrano. With all my expertise in the chemical process area, control was all new for me. Linear algebra-based control (LABC) was the first tool I used to control processes that should follow variable profiles. In a very short time, I was already controlling simple systems that were becoming more complex over time. Their successful application to laboratory-scale reactors, whose control was really difficult with other control techniques, was the next step. Only those who have lost hours of work because they cannot control their reactive system can understand this great satisfaction.

This simple and robust technique showed me that for something to work well it is not necessary to have a high degree of complexity. In general, and particularly in the academic field, there is a tendency to think that the more complex a technique is, the better it should work. And when comparing the results I obtained by using this controller with those from more complex control schemes, I verified that this is not always the case.

This book presents an important contribution to the field of process control, which implies a different way of facing the problem. It provides the reader with a clear grounding in LABC and is appropriate for those with a basic knowledge of classical control theory. It includes an explanation of the methodology and a broad range of well-worked-out application studies. Experimental case studies, which present the results of linear algebra-based controller implementations, are used to illustrate its successful practical application.

MSc. Ing. María Fabiana Sardella
Profesora Titular Facultad de Ingeniería
Universidad Nacional de San Juan – Argentina

Preface

The trajectory tracking problem is a very important one in control theory. The main goal is that some system variables follow a given evolution in the predefined time. These reference signals are often obtained by means of some optimization procedure (for example, determining the feed profile of a reactor to maximize production) or they can be generated online through the references that human operators give to robots in rescue operations, recognition, or vigilance. Also, robots that transport loads between production lines and warehouses, and more recently the case of vehicles moving without human intervention through the cities, fall into this category.

This book presents a new methodology for the design of controllers for trajectory tracking, where the controller design problem is linked to that of solving a system of linear equations. In this way, it is possible to deal with a complex problem from a simpler point of view. Moreover, in general when a problem can be presented from a simpler point of view, it is easier to obtain conclusions about the behavior of the system under study, and thus to know what modifications are required to improve its performance. An important step in the design procedure is to analyze under which conditions the system of equations has an exact solution. This allows to determine the desired value of some of the state variables whose reference is not given and may momentarily take values not well a priori defined. For that reason, we have called them as sacrificed variables. The greatest contribution of this book is to outline a procedure to be followed to design the controller that ensures that the system follows the reference signals. The system can be linear, nonlinear, monovariable, or multivariable; the only condition is that it should be minimum phase and the model should be affine in the control.

This book is based on the research we have carried out since 2005, when the methodology based on linear algebra was applied for the first time to design the control of a mobile robot based on its cinematic model. Then the technique was applied to more complex systems such as ships, airplanes, quad rotor, and chemical processes, as shown in the publications listed as references. Other than the basic procedure, the modifications of the original algorithm to take into account the perturbations and the uncertainty in the model are also described.

The final structure of the book is based on the work we have done in our research group as well as on the courses and seminars taught in different universities. The book can be used to introduce control of trajectory tracking as part of an advanced control course for undergraduates. On the whole, it can be used for a postgraduate course on control of trajectory tracking. The book has a practical orientation and is also suitable for process engineers.

San Juan, Argentina — Gustavo Scaglia
San Juan, Argentina — Mario Emanuel Serrano
Valencia, Spain — Pedro Albertos

Acknowledgments

The authors would like to thank a number of people who in various ways have made this book possible. Firstly, we thank Fabiana Sardella for her valuable collaboration in the preparation of this manuscript. Our thanks also go to Dra. Cecilia Fernández, Dra. Nadia Pantano, Dr. Leandro Rodríguez, Dr. Sebastián Godoy, Dr. Santiago Romoli, and Dr. Francisco Rossomando for their help in preparing some examples and revising the manuscript. Special thanks go to Jorge Romero, Rafael Fava, and Eduardo Strazza for their constant support and incentive.

The authors would like to thank Dr. Oscar Camacho and Ing. Marcos Herrera of the National Polytechnic School of Ecuador and Dr. Olga Lucía Quintero Montoya of the EAFIT University, Colombia, for their collaboration to develop several works.

Our thanks also go to our colleagues and friends from the Instituto de Ingeniería Química IIQ, especially to Ing. Pablo Aballay and Dr. Oscar Ortiz, as well as from the Departamento de Física de la Facultad de Ingeniería de la Universidad Nacional de San Juan.

Part of the material included in the book is the result of research work funded by CONICET (Consejo Nacional de Investigaciones Científicas y Técnicas de Argentina), the National University of San Juan Argentine (UNSJ), Instituto Universitario de Automática e Informática Industrial, Universitat Politècnica de València, Spain, and Prometeo, the special research Program funded by the Senescyt (Ecuador) providing the environmental conditions for a fruitful collaboration among the authors. Special thanks to our colleague Dr. Andrés Rosales (Escuela Politécnica Nacional (EPN), Quito) who participated in some of the initial works. We gratefully acknowledge these institutions for their support. Also, we thank the IIQ and the Instituto de Automatica (INAUT) from Facultad de Ingeniería from UNSJ for providing us the facilities and equipment necessary to develop the experimental tests.

Finally, all authors thank their families for their support, patience, and understanding of family time lost during the writing of the book.

Contents

Chapter 1
Introduction to Tracking Control

The main goal of a control system is to provide the appropriate input signals to the plant, in order to get a desired behavior of the controlled plant. Of course, the plant should be able to behave in the required behavior, and thence, the control system is generating the inputs among those of the admissible set. The desired behavior could be to remain in a stable condition, *the regulation problem*, or to follow a given trajectory, *the tracking problem*, in spite of external disturbances, starting from unknown initial conditions or under changes in the controlled plant parameters or structure. Although the control goal is different, both problems involve similar solutions and control design procedures. Perhaps the major difference is the existence of a desired reference trajectory, usually known in advance, that in the regulation problem can be considered as a constant. But stability, disturbance rejection, and steady-state behavior issues are common in both settings.

In a more formal way, the control problem can be stated as follows: given a plant where some information is measured through the output variables $y(t)$, determine the input signal $u(t)$ to fulfill the control goal, in spite of possible disturbances. Usually, the input signals are subjected to some constraints, and a model of the plant behavior, $y(t) = \mathbf{M}[u(t)]$, is available.

Most practical systems have a nonlinear behavior, and the solution of the control problem leads to suitable nonlinear controllers. Although the theory and control design procedures for linear systems is well established and accepted by the control community, there is a large variety of approaches to analyze the behavior and control design for nonlinear systems, and there is not a unique approach valid for all kinds of nonlinear systems. There are many ad hoc solutions only applicable for a kind of nonlinearity, and the general approaches (such as feedback linearization, sliding mode control, or model predictive control, among many others) are only valid or more suitable for some classes of nonlinearities.

In this monograph, a new approach to design the tracking control for a special class of nonlinear systems is presented. Although initially it was conceived to design the control for mobile robots, it has been proven to be very effective in the control of a variety of processes, including chemical processes, where the main goal is to

G. Scaglia et al., *Linear Algebra Based Controllers*,
https://doi.org/10.1007/978-3-030-42818-1_1

follow a given reference for some process variables. The design procedure and its application to different nonlinear plants are developed in the following chapters.

1.1 Tracking Control Problems

Although a lot of research on the stability analysis of dynamic controlled plants has been reported in the literature (Isidori, 1995; Khalil & Praly, 2014; Kothare & Morari, 1999), in the past decades, there has been significant interest in the trajectory tracking control. One of the main reasons is the appearance of autonomous vehicles, able to navigate in uncertain environments, following a trajectory, and avoiding possible obstacles (Antonelli, Chiaverini, Finotello, & Schiavon, 2001; Berglund, Brodnik, Jonsson, Staffanson, & Soderkvist, 2009; Yoon, Shin, Kim, Park, & Sastry, 2009). Thus, other than the stability issues, the control design should cope with disturbance rejection and the treatment of unknown initial conditions.

Moreover, the control of autonomous systems may be considered at different levels. From the simplest case of warning under unexpected conditions, just firing some alarms to prevent the change of the control system, to the case of fully independent systems where the control system must incorporate decision capabilities to deal with any change, either in the environment or inside the own system.

Two main control structures can be distinguished: feedforward control, when the control action is computed without actual information about the plant evolution, or feedback control, when the control action relies on the information gathered from the plant, in real time. Both approaches are complementary, and they should be jointly used to design the control solution. This is mainly the case when facing a tracking control problem, where information about the references to be tracked is available.

1.1.1 Feedforward Control

Consider the external representation of a linear system such as

$$y(s) = G(s)u(s) \tag{1.1}$$

where s is the Laplace complex variable and $G(s)$ is the transfer matrix. The ideal control to follow a desired output $y_r(s)$ will be

$$u(s) = G^{-1}(s)y_r(s) \tag{1.2}$$

This requires $G^{-1}(s)$ being feasible and stable. Perfect tracking will be achieved if there are no disturbances, the model matches the real plant, and the initial conditions are also matched.

For a nonlinear process, consider the mathematical representation of the process such as

$$y(t) = \mathbf{M}[u(t)] \quad (1.3)$$

where $\mathbf{M}[\bullet]$ is the mathematical operator, transforming the input signal into the output signal. If the model is perfect, and this operator is invertible, the control input required to track a desired output reference $y_r(t)$ would be generated by

$$u(t) = \mathbf{M}^{-1}[y_r(t)] \quad (1.4)$$

Thus, a feedforward control will achieve the reference tracking. But there are several reasons making this control inappropriate and insufficient: the model is inaccurate, and there are uncertainties in the model parameters, the initial output is not "matched" to the desired trajectory, some disturbances and some unmodeled dynamics may appear, and they are not considered in the model, and even the inverse operator $\mathbf{M}^{-1}[\cdot]$ could be unstable or unrealizable.

If the output model is implicit and the internal variables are defined in the desired reference, the inverse dynamics controller generating the control input acting on the plant allows for a simpler realization. A common example in robotics control literature can illustrate this principle. Assume a robot arm modeled by

$$M(q)\ddot{q} + f(q,\dot{q}) = u \quad (1.5)$$

where q are the generalized coordinates, $M(q)$ is the inertial matrix, f is a nonlinear vector function, involving the Coriolis, gravitational and friction terms, and u is the torque vector input, and there is a double differentiable desired reference $q_r(t)$. The tracking control can be generated by

$$u = M(q_r)\ddot{q}_r + f(q_r,\dot{q}_r) \quad (1.6)$$

But, again, no disturbances are allowed, the model parameters should be perfectly known, and the initial conditions of the robot arm should be matched by the controller: $q_r(0) = q(0); \dot{q}_r(0) = \dot{q}(0)$.

Due to these drawbacks, feedforward control cannot be applied as it is and, in any case, it should be complemented with some feedback control coping with disturbances and model uncertainties.

1.1.2 Feedback Control

It is well known that feedback control can stabilize unstable plants, and it can make the controlled plant robust against disturbances as well as model uncertainties. Many different approaches can be foreseen to design the feedback controller but, as already mentioned, they are very dependent on the kind of nonlinearity appearing in the model.

Among the many nonlinear feedback control design techniques, some of them are most common. *Feedback linearization* is aimed to pre-control the process to reduce it to a linear model. It requires the finding of a transformation of the nonlinear system into an equivalent linear system represented by a cascade of integrators through a change of variables and a suitable control input. This technique is very useful but not always applicable (Krener, 1999). In principle, the nonlinear system should be affine in the control.

Adaptive control is a very useful technique requiring a double loop. There are different approaches and a lot of literature describing all these options (Åström & Wittenmark, 2013; Landau, 1979). In general, the inner loop is a feedback control that may be designed for a local linear model of the plant, and the outer loop is an identifying loop trying to estimate the parameter variation in the plant model or to directly change ("adapt") the controller parameters. So, adaptive control mainly focuses on controlling plants under uncertain environments trying to improve the plant model and accordingly adjust on-line the control.

A different approach for the same control problem is the *robust control.* In this case, the uncertainty in the model or in the external disturbances (the environment) should be bounded, and the control is designed in such a way that satisfying performance is obtained under all the possible conditions (Morari & Zafiriou, 1989; Zhou & John, 1999). In the case of adaptive control, the changes are not bounded and even not known in advance.

A *cascade decomposition* can be foreseen for nonlinear process models with triangular structure, and then a backstepping control could be designed for lower triangular plants (Krstic, Protz, Paduano, & Kokotovic, 1995) and a forwarding control for upper triangular systems (Jankovic, Sepulchre, & Kokotovic, 1996).

Model Predictive Control allows dealing with nonlinear models, and it provides an optimal control signal by considering a limited horizon in a constrained environment. It usually involves a heavy computation load due to the optimization approaching, although very efficient real-time algorithms have been recently proposed to on-line apply this control (Camacho & Bordons, 2004; Rawlings, Meadows, & Muske, 1994).

If there are strong uncertainties and the model is not well known, *sliding mode control* is an appropriate control design methodology. In this approach, the controller applies strong control signals (not much dependent on the model) to drive the plant state to a sliding surface where a linearized control is applied (Utkin, 1993). Obviously, there is a commutation between control actions, depending on the plant state.

One critical feature in implementing feedback control is the information available to generate the control. In this sense, it should be distinguished between *output feedback control* and *state feedback control.* In the second case, the process state is available. That is, all the information about the process can be used and, if the process is controllable, its dynamics can be arbitrarily modified. On the other hand, if only the output signals are available, the control performance will be degraded even if a state observer is designed and implemented.

1.1.3 Iterative Tracking Control

In the feedforward control, a model of the process is required, and the performance of the control very much depends on the model accuracy. Sometimes, the process behavior is not well known or the model uncertainties are very strong. And the previous approach fails. The model requirement can be relaxed if the tracking activity is repetitive, that is, if the goal is to perform *repetitive motion tasks*.

In this more elaborated structure, the control is composed of two parts. One basic feedback control, which is model independent, such as a Proportional-Derivative (PD) controller, and a feedforward controller which is updated at each iteration being computed to minimize the tracking errors in the previous iteration and being reinitialized at the beginning of each trial. Thus, a learning capability is included in the controller design stage, reducing the tracking errors (Owens & Hätönen, 2005.)

In general, the design of a tracking controller will use some of the facilities of all these approaches and, as a general conclusion, it will be strongly conditioned by the available knowledge of the plant, that is, its model.

1.2 Process Model

In the following chapters, a control design methodology for nonlinear processes based on linear algebra is presented. As it will be shown, the control design is approached by process model inversion but in a soft way. This implies that the process model should be affine in the control.

One of the key ideas is that in tracking control, not all the process variables should follow a predefined reference. Thence, only the tracking variables will be forced to follow the references, and the rest of variables, denoted as *sacrificed variables*, are required to follow some ad hoc references, allowing a smooth behavior of the tracked variables. The reference will be assumed to be known, as well as its derivatives if so required.

By using the internal representation, the model will be as

$$\dot{x}(t) = F(x(t), u(t), d(t), t) \tag{1.7}$$

where $x \in R^r$ denotes the state of the system, $u \in R^m$ denotes the input, $t \in R$ denotes the continuous time (CT), and $d \in R^r$ denotes an external disturbance. If the model is time invariant, t will not be an entry in the function F.

Usually, the number of tracked variables is equal to or less than the number of inputs, that is, there is a maximum number of r-m sacrificed variables.

In the approach to be developed, the process model is assumed to be affine in the control, and the disturbances are assumed as additive ones. That is, (1.7) can be written as

$$\dot{x}(t) = f(x(t)) + g(x(t))u(t) + d(t) \tag{1.8}$$

where $f(\cdot)$ is an r-dimensional vector and $g(\cdot)$ is an $r \times m$ matrix.

Most of the time, the controller will be digitally implemented. Thus, either the controller model should be given in discrete time (DT) form being the result of the discretization of a CT controller or the controller has been directly designed for the discretized plant model.

A crucial issue in DT control is the selection of the sampling period. As usual, it should always be a compromise between the requirements of the computational load and the tracking accuracy.

1.3 Processes

Tracking control traditionally arises in the design of *servosystems*. This is very common in mechanical plants were the main control objective is to guarantee that a powered signal follows the reference provided by a signal in the information area. Typical examples are the displacement of a heavy object or the master–slave connection in some devices.

In this book, several examples of different processes are considered to design its control. In all cases, the goal is to track a reference in spite of model uncertainties and external disturbances. The manipulating robot is the paradigmatic example, as the end effector of the robot arm is required to perform different activities following a prescribed trajectory. But nowadays, the extensive use of autonomous vehicles has generalized the need of accurate tracking controllers acting on very different environments.

Surface, aerial, or aquatic vehicles will be the main goal of the book but also many other processes, like chemical reactors, will be considered as the control object. In any case, the tracking of a reference or profile will be the main requirement to show an appropriate plant behavior.

1.4 Outline of the Book

The book deals with a new technique to design the tracking control for a variety of processes. It is grounded on a mathematical model of the plant, and its development involves many concepts from systems theory and liner algebra. Thus, most of the required concepts from these disciplines are summarized in an appendix, leading the reader to look for a wider background in the literature.

In Chap. 2, the newly proposed methodology, the so-called Linear Algebra-Based (LAB) Control Design (CD) methodology, will be described and a procedure to design a tracking controller for a general process will be presented. The main decision steps are emphasized, and the expected difficulties are pointed out.

In Chap. 3, the LAB CD methodology is applied for a mobile robot, by using a simple kinematic model and assuming a perfectly known model and environment. The main properties of the designed control are shown, and the design steps are illustrated.

Chapter 4 includes a variety of tracking control solutions for the same process (a surface mobile robot) but considering different alternatives, including more accurate models, digital control, or disturbances.

Aquatic and aerial autonomous vehicles are the subject of Chap. 5. As the model becomes more complicated, more state variables should be considered, and the methodology to deal with sacrificed variables is illustrated. A simulation model for a marine vessel is provided, allowing the reader for a personal training, comparing the theoretical results with those obtained by simulation. Of course, the reader may develop similar diagrams to evaluate the different processes used through all the chapters to illustrate the key points of this new control design technique.

The applicability of the LAB CD methodology to control the evolution of chemical processes according to a required profile is discussed in Chap. 6. Also, the advantages of using simplified models experimentally obtained are discussed, and a general approach to model and control design for a first-order plus time delay (FOPTD) model of a nonlinear plant is presented.

Linked to the use of simplified models is the robustness of the controlled plant to changes in the model parameters. This is the subject of Chap. 7, where a restricted class of model uncertainties and disturbances is considered, leading to the classical integral actions, well recognized by the end users.

Finally, in Chap. 8, the main issues appearing in the application of this methodology are reviewed, and general guidelines are provided.

Through the book, many linear algebra concepts are used. To make easier the reading of the book, an appendix summarizing the main concepts on linear algebra and system dynamics is included at the end.

References

Antonelli, G., Chiaverini, S., Finotello, R., & Schiavon, R. (2001). Real-time path planning and obstacle avoidance for RAIS: An autonomous underwater vehicle. *IEEE Journal of Oceanic Engineering, 26*(2), 216–227.

Åström, K. J., & Wittenmark, B. (2013). *Adaptive control*. New York: Courier Corporation.

Berglund, T., Brodnik, A., Jonsson, H., Staffanson, M., & Soderkvist, I. (2009). Planning smooth and obstacle-avoiding B-spline paths for autonomous mining vehicles. *IEEE Transactions on Automation Science and Engineering, 7*(1), 167–172.

Camacho, E. F., & Bordons, C. (2004). *Model predictive control* (pp. 185–197). Berlin: Springer Verlag.

Isidori, A. (1995). *Nonlinear control systems* (3rd ed.). Berlin: Springer-Verlag.

Jankovic, M., Sepulchre, R., & Kokotovic, P. V. (1996). Constructive Lyapunov stabilization of nonlinear cascade systems. *IEEE Transactions on Automatic Control, 41*(12), 1723–1735.

Khalil, H. K., & Praly, L. (2014). High-gain observers in nonlinear feedback control. *International Journal of Robust and Nonlinear Control, 24*(6), 993–1015.

Kothare, M. V., & Morari, M. (1999). Multiplier theory for stability analysis of anti-windup control systems. *Automatica, 35*(5), 917–928.

Krener, A. J. (1999). Feedback linearization. In J. Baillieul & J. C. Willems (Eds.), *Mathematical control theory*. New York, NY: Springer.

Krstic, M., Protz, J. M., Paduano, J. D., & Kokotovic, P. V. (1995, December). Backstepping designs for jet engine stall and surge control. In *Proceedings of 1995 34th IEEE Conference on Decision and Control* (Vol. 3, pp. 3049–3055). New York: IEEE.

Landau, I. D. (1979). *Adaptive control: The model reference approach*. New York: Marcel Dekker.

Morari, M., & Zafiriou, E. (1989). *Robust Process Control*. Prentice Hall. Upper Saddle River.

Owens, D. H., & Hätönen, J. (2005). Iterative learning control—An optimization paradigm. *Annual Reviews in Control, 29*(1), 57–70.

Rawlings J. B., Meadows E. S., Muske K. R., (1994). "Nonlinear Model Predictive Control: A tutorial and Survey", *IFAC Symposium on Advance Control of Chemical Processes*, Kyoto, Japan, 25–27 May 1994, IFAC Proceedings Volumes, Vol. 27(2).

Utkin, V. I. (1993). Sliding mode control design principles and applications to electric drives. *IEEE Transactions on Industrial Electronics, 40*(1), 23–36.

Yoon, Y., Shin, J., Kim, H. J., Park, Y., & Sastry, S. (2009). Model-predictive active steering and obstacle avoidance for autonomous ground vehicles. *Control Engineering Practice, 17*(7), 741–750.

Zhou, K., & John, D. C. (1999). *Essentials of robust control*. Upper Saddle River: Prentice Hall.

Chapter 2
Control Design Technique

In this book, the tracking control problem for nonlinear systems is considered. The system under consideration may be any autonomous vehicle (ground, aerial, or marine) or any process where some references should be followed, like batch chemical reactors, bioreactors, or kilns.

Trajectory tracking control should provide the direction and speed of changes in the plant to guide the controlled plant along a pre-defined path. This desired path is the result of a higher-level decision system, providing the path planning based on some required strategies. In the case of autonomous vehicles, the control variables are the steering angle and the vehicle speed.

When dealing with trajectory tracking, three different scenarios can be considered: (1) to drive the plant from an initial position to a final one (the target). This is called *point-to-point* motion, and the actual trajectory as well as the time used to reach the target is not so relevant. (2) To drive the plant along a given geometrical trajectory, regardless of the timing. This is referred as *path following*, and the main concern is to follow the desired trajectory without large errors. (3) To drive the plant from an initial point along a required trajectory, following a prescribed timing. This is referred as *path tracking*, and it is the most complete trajectory control, implying the control of the geometrical evolution of the plant as well as its speed.

Path tracking is the main concern of this book, and a new methodology to design the appropriate controllers will be developed.

In this chapter, the main features of the newly proposed control design methodology are presented. First, the class of systems to be considered is defined and the key points of the design approach are drafted. Then, an introductory example allows the understanding of the main properties and to state the procedure to design the control. The Linear Algebra-Based Control Design (LAB CD) methodology is then outlined, and the properties of the controlled plant are analyzed. Its applicability in discrete time (DT) is shown to be immediate, and the treatment of uncertainties and disturbances is introduced. Finally, a summary of the LAB CD approach is given as a guideline for its application to a variety of processes in the following chapters.

G. Scaglia et al., *Linear Algebra Based Controllers*,
https://doi.org/10.1007/978-3-030-42818-1_2

2.1 Problem Statement

A mathematical model of the plant, as well as the trajectory to be followed, is assumed to be given. The basic general model is expressed in state space form, as

$$\begin{aligned} \dot{x}(t) &= F(x(t), u(t), d(t), t) \\ y(t) &= H(x(t), u(t), t) \end{aligned} \tag{2.1}$$

where $x \in R^r$ denotes the state of the system, $u \in R^m$ denotes the input, $y \in R^p$ denotes the output of the system, and $d \in R^r$ denotes an external disturbance. The study of this control problem has deserved a lot of research in the literature, and the best solution is based, in many cases, on an ad hoc approach for a particular situation.

In our study, some simplifications of the model (2.1) are assumed:

1. The model is affine in the control.
2. The model is minimum phase.
3. The model is time invariant.
4. The state is measurable.
5. The model is exact, and there are no disturbances.
   That is, the initial model is given by

$$\begin{aligned} \dot{x}(t) &= f(x(t)) + g(x(t))u(t) \\ y(t) &= x(t) \end{aligned} \tag{2.2}$$

The fifth assumption will be relaxed later on, and some uncertainties and disturbances will be considered. Also, the fourth assumption can be relaxed if observers or dynamic output feedback is implemented. Assumption 3 can also be removed if the parameter variation law is known, with additional complexity in the control computation. Assumption 1 is not very restrictive as most of the processes to be considered present this control structure. Finally, Assumption 2 is required due to some model cancellation in the formal solution.

To fully describe the tracking problem, some reference trajectories should be provided. This is called the path tracking problem, and it will be assumed that the motion planner provides a feasible trajectory; that is, a trajectory that can be followed by applying appropriate control input. Thus, an additional assumption is:

6. The reference as well as its derivatives is known.

But in many cases, the desired trajectory is only defined for some of the state variables (usually positions and/or velocities), and the trajectory of the remaining state variables is optional, allowing for some freedom in implementing the control.

There are many approaches in the literature (Bouhenchir, Cabassud, & Le Lann, 2006; Chwa, 2004; Fukao, Nakagawa, & Adachi, 2000; Kanayama, Kimura, Miyazaki, & Noguchi, 1990) to deal with this problem. Some solutions are simple, but they require a simplified linear model of the plant, requiring adaptation or other strategies for complex models. Some other solutions use a complex model of the plant and require heavy computation, not being suitable for on-line applications. In our case, by exploiting the optionality in the evolution of some state variables, the control problem will be formulated in an algebraic scenario, leading to an easy to compute control (Scaglia, Montoya, Mut, & Di Sciascio, 2009).

In this chapter, the basic issues as well as their solutions are presented, the concrete results being illustrated in the following chapters.

2.2 Control Design

Let us split the state vector into two parts: the tracked variables $\xi(t) \in R^{r_1}$ and the remaining variables, $z(t) \in R^{r-r_1}$. These variables will be denoted as *sacrificed* variables. Usually, the number of variables to be tracked (r_1) is equal to the number of independent control actions (m), and the number of sacrificed variables depends on the plant model.

$$\begin{bmatrix} \dot{\xi}(t) \\ \dot{z}(t) \end{bmatrix} = \begin{bmatrix} f_\xi(\xi(t), z(t)) \\ f_z(\xi(t), z(t)) \end{bmatrix} + \begin{bmatrix} g_\xi(\xi(t), z(t)) \\ g_z(\xi(t), z(t)) \end{bmatrix} u(t) \tag{2.3}$$

$\xi_{\text{ref}}(t)$ is provided by the motion planner, and $z_{\text{ref}}(t)$ will be defined later on. So, the control problem can be stated as: given the model (2.3) and the references $\xi_{\text{ref}}(t)$ find the control input $u(t)$ forcing the substate vector $\xi(t)$ to follow the reference $\xi_{\text{ref}}(t)$ satisfying the model. The derivative of the state variables required to be tracked, $\xi(t)$, are initially replaced by that of the reference variables, assuming a smooth approaching, for instance, proportional to the error. In the problem statement, there is no reference for the sacrificed variables. Nevertheless, the derivative of some sacrificed variables, if so required, will be defined later on, and a similar approaching will be considered. That is,

$$\begin{bmatrix} \dot{\xi}(t) \\ \dot{z}(t) \end{bmatrix} = \begin{bmatrix} \dot{\xi}_{\text{ref}}(t) + k_\xi[\xi_{\text{ref}}(t) - \xi(t)] \\ \dot{z}_{\text{ref}}(t) + k_z[z_{\text{ref}}(t) - z(t)] \end{bmatrix} \tag{2.4}$$

where k_ξ, k_z are two diagonal matrices (dimension r_1, $r - r_1$, respectively), which are the control parameters. Combining (2.3) and (2.4), the following model of the controlled plant is obtained

$$\begin{bmatrix} \dot{\xi}_{\text{ref}}(t) + k_\xi[\xi_{\text{ref}}(t) - \xi(t)] - f_\xi(\xi(t), z(t)) \\ \dot{z}_{\text{ref}}(t) + k_z[z_{\text{ref}}(t) - z(t)] - f_z(\xi(t), z(t)) \end{bmatrix} = \begin{bmatrix} g_\xi(\xi(t), z(t)) \\ g_z(\xi(t), z(t)) \end{bmatrix} u(t) \tag{2.5}$$

So, the main features of the Linear Algebra-Based Control Design (LAB CD) approach rely on both, the selection of these parameters (or those related to an approaching more elaborated than that in (2.4)) and the searching of a solution of (2.5) to derive the control action, $u(t)$.

The above equation can be rewritten as

$$b(t) = A(t)u(t) \tag{2.6}$$

where A is a known $r \times m$-dimensional matrix and b is an r-dimensional vector, some of whose entries are partially unknown, $(z_{ref}(t), \dot{z}_{ref}(t))$.

In order to find a solution for $u(t)$ in (2.5), it is required that b must be a linear combination of the column vectors of A. This will determine the possible value for the reference of some sacrificed variables, $z_{ref}(t)$, as well as modify the first raw of (2.4) in such a way that the first raw of (2.5) accomplishes this condition. Once b and A are defined, the control will be obtained by solving (2.6), using the least square solution

$$u(t) = A^{\dagger}(t)b(t) \tag{2.7}$$

where $A^{\dagger}(t)$ stands for the pseudoinverse matrix of $A(t)$.

2.3 An Introductory Example

Let us consider a simple model of an XY plotter. These devices are very common in industry not only for recording purposes but also as cutters or performing other activities, carrying the appropriate end effector. This end effector is denoted as gantry, and it can support a pen, a cutter, or similar.

In essence, an XY plotter has a movement in the XY plan. The X displacement is done at a constant speed, and the Y displacement is achieved by acting on the gantry by means of a DC motor.

A simplified model of this plotter is given by

$$m\ddot{y}(t) + r\dot{y}(t) + ky(t) = f(t) \tag{2.8}$$

where m is the moving mass of the gantry, r is the friction coefficient, k is the spring constant, and $f(t)$ is the force provided by the motor to move the gantry.

The tracking control problem can be stated as follows: given a plotter with a model (2.8), generate the control action $f(t)$ to track a reference signal $y_{ref}(t)$. The model (2.8) can be transformed into the internal representation (2.3), yielding

$$\begin{bmatrix} \dot{\xi}(t) \\ \dot{z}(t) \end{bmatrix} = \begin{bmatrix} 0 & 1 \\ -k/m & -r/m \end{bmatrix} \begin{bmatrix} \xi(t) \\ z(t) \end{bmatrix} + \begin{bmatrix} 0 \\ 1/m \end{bmatrix} u(t) \tag{2.9}$$

where $\xi(t) = y(t)$, $z(t) = \dot{y}(t)$, and $u(t) = f(t)$.

The reference for the position is assumed to be known, $y_{\text{ref}}(t)$. Thus, equation (2.5) can be written as

$$\begin{bmatrix} \dot{\xi}_{\text{ref}}(t) + k_\xi[\xi_{\text{ref}}(t) - \xi(t)] - z(t) \\ \dot{z}_{\text{ref}}(t) + k_z[z_{\text{ref}}(t) - z(t)] + {}^{k}\!/_{m}\xi(t) + {}^{r}\!/_{m}z(t) \end{bmatrix} = \begin{bmatrix} 0 \\ {}^{1}\!/_{m} \end{bmatrix} u(t) \tag{2.10}$$

In this case, due to the elements in A, the first entry of b should be null. That is, the reference for the sacrificed variable $z(t)$ should be such that

$$\dot{\xi}_{\text{ref}}(t) + k_\xi[\xi_{\text{ref}}(t) - \xi(t)] - z_{\text{ref}}(t) = 0 \tag{2.11}$$

to make zero the first entry of b. Note that the second entry is not modified.

The derivative of the output, according to (2.9), should be $\dot{\xi}(t) = z(t)$. So, adding (2.11) in the right-hand side, it yields

$$\dot{\xi}(t) = \dot{\xi}_{\text{ref}}(t) + k_\xi[\xi_{\text{ref}}(t) - \xi(t)] - z_{\text{ref}}(t) + z(t) \tag{2.12}$$

where $z_{\text{ref}}(t) - z(t) = e_z(t)$ is the tracking error of the sacrificed variable. Thus, the tracking error will be

$$\dot{\xi}_{\text{ref}}(t) - \dot{\xi}(\text{t}) = -k_\xi[\xi_{\text{ref}}(t) - \xi(t)] + e_z(t) \Rightarrow \dot{e}_\xi = -k_\xi e_\xi(t) + e_z(t) \tag{2.13}$$

Substituting (2.11) in (2.10), and pre-multiplying by A^T(where $()^T$ stands for transposition), the control action should be

$$u(t) = m\left[\dot{z}_{\text{ref}}(t) + k_z\left(\dot{\xi}_{\text{ref}}(t) + k_\xi[\xi_{\text{ref}}(t) - \xi(t)] - z(t)\right) + {}^{k}\!/_{m}\xi(t) + {}^{r}\!/_{m}z(t)\right] \tag{2.14}$$

Observe that, in this simple case, the control action can be directly computed from the second raw in (2.10). $\dot{z}_{\text{ref}}(t)$ is required to compute the control action. It can be evaluated if the derivative in (2.11) is computable. This will lead to the control action, as a function of the plant variables ($\xi(t)$ and $z(t) = \dot{\xi}(t)$) and the reference

$$\begin{aligned} u(t) &= m\ddot{\xi}_{\text{ref}}(t) + m(k_\xi + k_z)\dot{\xi}_{\text{ref}}(t) + mk_\xi k_z\xi_{\text{ref}}(t) \\ &-(mk_\xi k_z - k)\xi(t) - (mk_\xi + mk_z - r)\dot{\xi}(t); \\ u(t) &= u_f(t) + u_b(t) \end{aligned} \tag{2.15}$$

The control law (2.15) can be rewritten as composed by two terms: a dynamic feedforward of the position reference, u_f, and a state feedback

$$u_b(t) = (k - mk_\xi k_z)\xi(t) + (r - mk_\xi - mk_z)z(t) \tag{2.16}$$

Other than the feedforward action, the dynamics of the controlled plant will be determined by that of the plant (2.9) with the feedback (2.16), that is:

$$\begin{bmatrix} \dot{\xi}(t) \\ \dot{z}(t) \end{bmatrix} = \begin{bmatrix} 0 & 1 \\ -k_\xi k_z & -(k_\xi + k_z) \end{bmatrix} \begin{bmatrix} \xi(t) \\ z(t) \end{bmatrix} \tag{2.17}$$

Thus, the controlled plant eigenvalues are $\lambda \in \{-k_\xi, -k_z\}$. They should be negative to guarantee the stability of the closed loop. Thus, the control parameters should be positive.

At this moment, it is interesting to analyze the behavior of the tracking error, to be used later on in more complicated plants. This error was defined as

$$\begin{bmatrix} e_\xi \\ e_z \end{bmatrix} = \begin{bmatrix} \xi_{\text{ref}} - \xi \\ z_{\text{ref}} - z \end{bmatrix} \tag{2.18}$$

where the argument (t) has been omitted for simplicity in the notation. Although from (2.4),

$$\dot{\xi} = \dot{\xi}_{\text{ref}} + k_\xi(\xi_{\text{ref}} - \xi)$$

and due to the constraint in (2.5), the tracking error is computed as (2.13)

$$\begin{bmatrix} \dot{e}_\xi(t) \\ \dot{e}_z(t) \end{bmatrix} = \begin{bmatrix} -k_\xi[\xi_{\text{ref}}(t) - \xi(t)] + z_{\text{ref}}(t) - z(t) \\ -k_z[z_{\text{ref}}(t) - z(t)] \end{bmatrix} \tag{2.19}$$

So,

$$\begin{bmatrix} \dot{e}_\xi \\ \dot{e}_z \end{bmatrix} = \begin{bmatrix} -k_\xi & 1 \\ 0 & -k_z \end{bmatrix} \begin{bmatrix} e_\xi \\ e_z \end{bmatrix} \tag{2.20}$$

Remark: This equation points out that the initial action of the LAB controller is to asymptotically reduce to zero the error of the sacrificed variable, $e_z(t) = e_z(0)\mathrm{e}^{-k_z t}$, and later on reduce the error of the tracked variables, $\dot{e}_\xi(t) = -k_\xi e_\xi(t) + e_z(0)\mathrm{e}^{-k_z t}$.

Remark: If the plant model would be a second-order nonlinear plant as described by

$$\ddot{y}(t) = f(y(t), \dot{y}(t)) + g(y(t), \dot{y}(t))u$$

instead of (2.9), the internal representation could be expressed as

$$\begin{bmatrix} \dot{\xi}(t) \\ \dot{z}(t) \end{bmatrix} = \begin{bmatrix} z(t) \\ f(\xi(t), z(t)) \end{bmatrix} + \begin{bmatrix} 0 \\ g(\xi(t), z(t)) \end{bmatrix} u(t)$$

and following the same reasoning than before, the nonlinear control feedback will be

$$u = \frac{1}{g(\xi(t), z(t))} [-f(\xi(t), z(t)) - k_\xi k_z \xi(t) - (mk_\xi + mk_z) z(t)]$$

and the error dynamics will also be represented by (2.20), leading to closed-loop eigenvalues similar to (2.17). Observe that the first term of the control action cancels part of the process model, requiring Assumption 2, in Sect. 2.1.

In this case, the feedback control law is precisely the one obtained if a feedback linearization approach were applied to this nonlinear model.

2.4 Linear Algebra-Based Control Design Methodology

Once the potential interest of this methodology has been pointed out in the introductory example, a formal procedure to apply the LAB CD methodology to a more general tracking problem is presented.

As previously said, LAB CD is mainly used to design the tracking control of a nonlinear plant whose model is affine in the control action. Initially, the remaining assumptions detailed in Sect. 2.1 are also assumed:

1. The model is minimum phase.
2. The plant model is time invariant.
3. There are no disturbances, external or internal, due to uncertainties in the model.
4. The state is accessible and measured.
5. The references to be tracked, as well as its derivatives (at least the first- and second-order derivatives), are known and accessible.

Some of these assumptions will be relaxed later on.

In summary, the following steps should be followed:

Step 1. Obtain an internal representation of the plant model, as in (2.2)

$$\dot{x}(t) = f(x(t)) + g(x(t))u(t)$$

Step 2. Split the state vector into two subvectors, collecting the state variables to be tracked, $\xi(t)$, whose reference is given, and the remaining state variables, $z(t)$, denoted as sacrificed variables, whose reference will be determined. The new model is like (2.5)

$$\begin{bmatrix} \dot{\xi}(t) \\ \dot{z}(t) \end{bmatrix} = \begin{bmatrix} f_\xi(\xi(t), z(t)) \\ f_z(\xi(t), z(t)) \end{bmatrix} + \begin{bmatrix} g_\xi(\xi(t), z(t)) \\ g_z(\xi(t), z(t)) \end{bmatrix} u(t)$$

Step 3. Define the state variables' derivatives as an approximation of those of their references. This approximation is the control design stage. For any of the state variables, the difference between their derivative and the corresponding reference is a function of the current error between the state and its reference. In the simplest case, this function is just a proportional factor, such as expressed in (2.4)

$$\begin{bmatrix} \dot{\xi}(t) \\ \dot{z}(t) \end{bmatrix} = \begin{bmatrix} \dot{\xi}_{\text{ref}}(t) + k_\xi[\xi_{\text{ref}}(t) - \xi(t)] \\ \dot{z}_{\text{ref}}(t) + k_z[z_{\text{ref}}(t) - z(t)] \end{bmatrix}$$

where k_ξ, k_z are diagonal matrices whose elements weight the tracking error in following the references. It is worth to remind that not all the sacrificed variables need to have a reference—their dynamics being determined by the computed control actions. The selection of these parameters as well as the weighting of the tracking errors will define the control law.

In this step, the control can be designed, assuming a more complicated approaching instead of (2.4), but the computation disadvantages exceed the possible benefits.

Step 4. Combining the last two equations, a model of the controlled plant (2.5) is obtained

$$\begin{bmatrix} \dot{\xi}_{\text{ref}}(t) + k_\xi(\xi_{\text{ref}}(t) - \xi(t)) - f_\xi(\xi(t), z(t)) \\ \dot{z}_{\text{ref}}(t) + k_z(z_{\text{ref}}(t) - z(t)) - f_z(\xi(t), z(t)) \end{bmatrix} = \begin{bmatrix} g_\xi(\xi(t), z(t)) \\ g_z(\xi(t), z(t)) \end{bmatrix} u(t)$$

which can be summarized as (2.6)

$$b(t) = A(t)u(t)$$

The $r \times m$ A-matrix is known from (2.2). The reference value of the sacrificed variables, as well as their derivatives, appearing in the r-dimensional vector b, is undefined. In order to have an exact solution for $u(t)$, b should be a linear combination of the column vectors of A. The reference of the required sacrificed variables, as well as their derivatives, is defined in such a way that this condition is fulfilled.

Remark: This computation may require some approximations to be discussed later on, when dealing with different applications.

Step 5. Solving the last equation in an optimal way, by using the least square approach, the control action is computed as

$$u(t) = A^{\dagger}(t)b(t)$$

Remark: As previously discussed, the control parameters are embedded in the vector $b(t)$, and their tuning can be done by ad hoc solutions, like in the simple example of the plotter, or by using general optimization approaches, like the Monte Carlo searching method, as shown in the following chapters.

2.5 The Tracking Error Equation

In the nonlinear case, the tracking error is much more involved than (2.19), but it can also be derived as follows: assume that the reference for all the sacrificed variables is required. In step 4, to make feasible the solution of (2.5), the reference should be such that

$$\begin{bmatrix} \dot{\xi}_{\text{ref}}(t) + k_\xi(\xi_{\text{ref}}(t) - \xi(t)) - f_\xi(\xi(t), z_{\text{ref}}(t)) \\ \dot{z}_{\text{ref}}(t) + k_z(z_{\text{ref}}(t) - z(t)) - f_z(\xi(t), z(t)) \end{bmatrix} = \begin{bmatrix} g_\xi(\xi(t), z_{\text{ref}}(t)) \\ g_z(\xi(t), z(t)) \end{bmatrix} u(t) \quad (2.21)$$

has a unique solution. Thus, subtracting (2.3), it yields

$$\begin{bmatrix} \dot{\xi}_{\text{ref}}(t) - \dot{\xi}(t) + k_\xi(\xi_{\text{ref}}(t) - \xi(t)) - [f_\xi(\xi(t), z_{\text{ref}}(t)) - f_\xi(\xi(t), z(t))] \\ \dot{z}_{\text{ref}}(t) - \dot{z}(t) + k_z(z_{\text{ref}}(t) - z(t)) \end{bmatrix}$$
$$= \begin{bmatrix} g_\xi(\xi(t), z_{\text{ref}}(t)) - g_\xi(\xi(t), z(t)) \\ 0 \end{bmatrix} u(t) \quad (2.22)$$

That is,

$$\begin{bmatrix} \dot{e}_\xi \\ \dot{e}_z \end{bmatrix} = \begin{bmatrix} -k_\xi e_\xi \\ -k_z e_z \end{bmatrix} + \begin{bmatrix} f_\xi(\xi, z_{\text{ref}}) - f_\xi(\xi, z) \\ 0 \end{bmatrix} - \begin{bmatrix} g_\xi(\xi, z_{\text{ref}}) - g_\xi(\xi, z) \\ 0 \end{bmatrix} u \quad (2.23)$$

$$\begin{bmatrix} \dot{e}_\xi \\ \dot{e}_z \end{bmatrix} = \begin{bmatrix} -k_\xi e_\xi \\ -k_z e_z \end{bmatrix} + \begin{bmatrix} h_\xi(\xi, z_{\text{ref}}, z, u) \\ 0 \end{bmatrix} \quad (2.24)$$

These additional terms will determine the evolution of the errors. In the applications developed in the following chapters, it will be shown that under some well-defined conditions these terms will vanish, and the stability of the controlled plant will be ensured as far as the controller parameters k_ξ and k_z are positive. Moreover, if not all the sacrificed variables should follow a predetermined reference, the second row in (2.21)–(2.24) will be different, as shown in some applications later on developed.

An especially interesting feature of the LAB CD approach shown in (2.24) is that, first, the tracking errors for the sacrificed variables are driven to zero, and afterwards the trajectory tracking errors are also driven to zero.

2.6 Linear Algebra-Based Control Design in Discrete Time

Nowadays, all the controllers are implemented in digital systems. Thus, the controller should be expressed in DT. One option is to discretize the previously computed control action, but better results can be obtained if the controller is derived for a DT model of the plant. Also, the initial model of the plant can be expressed in DT, and, thus, the control design should be done in this framework.

Let us assume a model of the plant as given by (2.2). A sampler of period T is applied to get the measurements, and a hold device with the same period is used to apply the control action. The first step now is to discretize this model. The simplest approach is the Euler approximation

$$\left.\frac{d[x(t)]}{dt}\right|_{t=nT} \simeq \frac{x((n+1)T)-x(nT)}{T} = \frac{x_{n+1}-x_n}{T} \tag{2.25}$$

where n is the sampling time instant. Many other approximations can be used, but if the sampling period is short enough, the Euler approximation is acceptable, and it can be easily applied to nonlinear plants.

So, in the DT framework, the control design steps will be similar than those outlined before, namely:

Step 1. Get an internal representation of the plant in DT. If the initial model is as given by (2.2), the DT model will be

$$x_{n+1} = x_n + T[f(x_n) + g(x_n)u_n] \tag{2.26}$$

Step 2. Split the state vector into two subvectors, collecting the state variables to be tracked, ξ_n, whose reference is given, and the remaining state variables, z_n, denoted as sacrificed variables, whose reference will be determined. The new model is

$$\begin{bmatrix} \xi_{n+1} \\ z_{n+1} \end{bmatrix} = \begin{bmatrix} \xi_n \\ z_n \end{bmatrix} + T\left(\begin{bmatrix} f_\xi(\xi_n, z_n) \\ f_z(\xi_n, z_n) \end{bmatrix} + \begin{bmatrix} g_\xi(\xi_n, z_n) \\ g_z(\xi_n, z_n) \end{bmatrix} u_n \right) \tag{2.27}$$

Step 3. Define the next value of the state variables as an approximation of those of their references. This approximation is the control design stage. For any of the state variables, their increment as well as that of the corresponding reference is a function of the current error between this state variable and its reference. In the simplest case, this function is just a proportional factor, such as expressed in (2.4). Now, in DT

$$\begin{bmatrix} \xi_{n+1} - \xi_n \\ z_{n+1} - z_n \end{bmatrix} = \begin{bmatrix} \xi_{\mathrm{ref},n+1} - \xi_{\mathrm{ref},n} + T\kappa_\xi\left[\xi_{\mathrm{ref},n} - \xi_n\right] \\ z_{\mathrm{ref},n+1} - z_{\mathrm{ref},n} + T\kappa_z\left[z_{\mathrm{ref},n} - z_n\right] \end{bmatrix} \quad (2.28)$$

where, as before, κ_ξ, κ_z are diagonal matrices whose elements weight the tracking error in following the references. In the following, let us denote by $k_\xi = I - T\kappa_\xi$; $k_z = I - T\kappa_z$. This allows to write (2.28) as

$$\begin{bmatrix} \xi_{n+1} \\ z_{n+1} \end{bmatrix} = \begin{bmatrix} \xi_{\mathrm{ref},n+1} \\ z_{\mathrm{ref},n+1} \end{bmatrix} - \begin{bmatrix} k_\xi\left(\xi_{\mathrm{ref},n} - \xi_n\right) \\ k_z\left(z_{\mathrm{ref},n} - z_n\right) \end{bmatrix} \quad (2.29)$$

In this equation, it is clear that for $k_\xi = k_z = I$ the tracking error remains constant, whereas for $k_\xi = k_z = 0$ the error is cancelled in one sampling period. Intermediate values of the coefficients will provide a smoother evolution decreasing the error progressively. These are the controller parameters to be tuned.

Step 4. Combining (2.27) and (2.29), a model of the controlled plant (2.5) is obtained as

$$\begin{bmatrix} \xi_{\mathrm{ref},n+1} - k_\xi\left(\xi_{\mathrm{ref},n} - \xi_n\right) - \xi_n - Tf_\xi(\xi_n, z_n) \\ z_{\mathrm{ref},n+1} - k_z\left(z_{\mathrm{ref},n} - z_n\right) - z_n - Tf_z(\xi_n, z_n) \end{bmatrix} = \begin{bmatrix} Tg_\xi(\xi_n, z_n) \\ Tg_z(\xi_n, z_n) \end{bmatrix} u_n \quad (2.30)$$

which can be summarized as

$$b_n = A_n u_n \quad (2.31)$$

The required value of $z_{\mathrm{ref},\ n}$ will be determined to ensure the solvability of (2.30).

Step 5. Once $z_{ref,\ n}$ has been evaluated, the value of $z_{ref,\ n+1}$ will be estimated, extrapolating the previously defined values of the sequence $\{z_{ref,\ n}\}$. Different approaches can be followed for this extrapolation, but simple ones will be shown to provide excellent results if the changes in the references are slow with respect to the sampling period. This will be illustrated in the applications developed in the following chapters.

Solving (2.31) by least squares, the control action will be obtained as

$$u_n = A_n^\dagger b_n \quad (2.32)$$

2.7 Linear Algebra-Based Control Design Under Uncertainties in the Model

As previously stated in (2.1), some disturbances $d(t)$ may modify the computed trajectory of the controlled plant. These disturbances may be internal, due to changes in the structure of the plant or in its parameters, or external, due to changes in the environment. The controllers should counteract or at least reduce the effects of these disturbances. The nature of this disturbance, as well as its entry in (2.1), could be very different, and all the approaches to counteract the effects of the disturbances rely on at least a partial knowledge of the disturbance and/or its accessibility. So, the effectiveness of a disturbance cancellation approach very much depends on the disturbance model and its impact on the plant.

In our setting, the model we are going to consider is (2.2), with an additive disturbance in the state. That is

$$\dot{x}(t) = f(x(t)) + g(x(t))u(t) + d(t) \tag{2.33}$$

Note that this disturbance may represent an external disturbance $w(t)$ as well as a model mismatch. If the actual plant is

$$\begin{aligned} &\dot{x}(t) = f_p(x(t)) + g_p(x(t))u(t) + w(t) \\ &\Rightarrow d(t) = \left[\, f_p(x(t)) - f(x(t))\right] + \left[g_p(x(t))u(t) - g(x(t))\right]u(t) + w(t) \end{aligned} \tag{2.34}$$

So, the disturbance in (2.33) is an unknown signal, depending on the model mismatch and external disturbances. Note that if the system is stable, the inputs are bounded, and the functions f and g are Lipschitz, then d will be a bounded uncertainty. As usual, several assumptions about this disturbance can be adopted. The most common is to assume that it is a polynomial function of time. In the simplest case, it is a constant, but it may be a ramp or any other power of time.

Let us first consider a constant disturbance, $d(t) = d_0$. The control action should cancel its steady-state effect. Thus, the derivative approaching as defined in (2.4) should include an integrative term such as

$$\begin{bmatrix} \dot{\xi}(t) \\ \dot{z}(t) \end{bmatrix} = \begin{bmatrix} \dot{\xi}_{\text{ref}}(t) + k_\xi[\xi_{\text{ref}}(t) - \xi(t)] + K_\xi U_\xi(t) \\ \dot{z}_{\text{ref}}(t) + k_z[z_{\text{ref}}(t) - z(t)] + K_z U_z(t) \end{bmatrix} \tag{2.35}$$

where K_ξ and K_z are tunable diagonal matrices and

$$U_\xi(t) = \int_0^t e_\xi(\tau)d\tau; \;\; U_z(t) = \int_0^t e_z(\tau)d\tau \tag{2.36}$$

For the disturbed plant (2.33), the tracking errors (2.24) will be

$$\begin{bmatrix} \dot{e}_\xi \\ \dot{e}_z \end{bmatrix} = -\begin{bmatrix} k_\xi e_\xi \\ k_z e_z \end{bmatrix} + \begin{bmatrix} h_\xi(\xi, z, u) \\ 0 \end{bmatrix} - \begin{bmatrix} K_\xi U_\xi \\ K_z U_z \end{bmatrix} + d \tag{2.37}$$

and taking the derivative in (2.37), it yields

$$
\begin{bmatrix} \ddot{e}_{\xi} + k_{\xi}\dot{e}_{\xi} + K_{\xi}e_{\xi} \\ \ddot{e}_{z} + k_{z}\dot{e}_{z} + K_{z}e_{z} \end{bmatrix} = \begin{bmatrix} \dot{h}_{\xi}(\xi, z, u) \\ 0 \end{bmatrix} \tag{2.38}
$$

If the right-hand term is bounded, the roots of the polynomials in the left-hand right terms will determine the stability of the controlled plant which is only dependent on the controller parameters.

The application of these integral terms, even more complicated if the character of the disturbance is assumed polynomial, will be illustrated in several applications in the following chapters. A detailed treatment of the uncertainty is included in Chap. 7.

2.8 Summary of Linear Algebra-Based Control Design Methodology

The main features of the proposed tracking control design methodology are:

1. Easy to apply for nonlinear plants with minor constraints in the model (affine in the control and minimum phase), assuming the state measurability.
2. The computation burden is reduced.
3. The computation of the references for the sacrificed variables may involve some difficulties.
4. Can be applied for CT or DT plant models.
5. The controller structure selection is simple.
6. The parameters tuning process is explicit.
7. A simple selection of these parameters guarantees the controlled plant stability.
8. The parameter tuning to get more exigent performance is not straightforward.
9. Uncertainty in the model can be easily handled, assuming some disturbance knowledge.

In the following chapters, the LAB CD methodology is applied to design the tracking control of different plants, from autonomous vehicles to chemical processes, illustrating the advantages and drawbacks of this approach.

In particular, the controller parameters selection based on optimality indices will be presented by using the Monte Carlo approach.

References

Bouhenchir, H., Cabassud, M., & Le Lann, M. V. (2006). Predictive functional control for the temperature control of a chemical batch reactor. *Computers & Chemical Engineering, 30*(6–7), 1141–1154.

Chwa, D. (2004). Sliding-mode tracking control of nonholonomic wheeled mobile robots in polar coordinates. *IEEE Transactions on Control Systems Technology, 12*(4), 637–644.

Fukao, T., Nakagawa, H., & Adachi, N. (2000). Adaptive tracking control of a nonholonomic mobile robot. *IEEE Transactions on Robotics and Automation, 16*(5), 609–615.

Kanayama, Y., Kimura, Y., Miyazaki, F., & Noguchi, T. (1990, May). A stable tracking control method for an autonomous mobile robot. In *Proceedings of the IEEE International Conference on Robotics and Automation* (pp. 384–389). New York: IEEE.

Scaglia, G., Montoya, L. Q., Mut, V., & Di Sciascio, F. (2009). Numerical methods based controller design for mobile robots. *Robotica, 27*(2), 269–279.

Chapter 3
Application to a Mobile Robot

The use of autonomous mobile robots in industry, agriculture, and homes has grown exponentially in recent years. Mobile robots are currently used to release from cleaning tasks; to access dangerous environments; to improve performance, quality, and achieve accurate applications in agriculture; for radioactive operations in nuclear scenarios, in order to minimize risks and where human operator presence is restricted or prohibited; etc.

To achieve high-precision trajectory tracking control for the wheeled mobile robot (WMR), many sophisticated control approaches have been proposed in the past. These methods can be characterized by two research paradigms, based on whether the WMR is described by a kinematic model or a dynamic model. Thus, the tracking-control problem is classified as either a kinematic or a dynamic tracking-control problem. Obviously, the kinematic design is simpler.

The mobile robot is the most frequently used device to deal with tracking problems. There is a lot of literature with different proposals (see, for instance, Li et al., 2015, Sun, Tang, Gao, & Zhao, 2016, Proaño, Capito, Rosales, & Camacho, 2015, Panahandeh, Alipour, Tarvirdizadeh, & Hadi, 2019, and many others). A comparative analysis will be reported in the next chapter, including the results obtained by the use of the new methodology.

In this chapter, the use of the Linear Algebra-Based Control Design methodology will be illustrated, being applied to a mobile robot (Scaglia, Montoya, Mut, & Di Sciascio, 2009; Scaglia, Quintero, Mut, & di Sciascio, 2008; Serrano, Godoy, Quintero, & Scaglia, 2017). Initially, the simplest kinematic continuous time model of the robot will be used, and no disturbances will be considered. The simplicity of the proposed control structure is based on the model, and a simulation diagram will allow the quick implementation of the control. A procedure to determine the controller parameters will be outlined, and the performance of the controlled plant, in both the transient (stability) and steady-state behaviors, will be analyzed (Scaglia, Serrano, Rosales, & Albertos, 2019).

G. Scaglia et al., *Linear Algebra Based Controllers*,
https://doi.org/10.1007/978-3-030-42818-1_3

3.1 Kinematic Control of a Mobile Robot

A nonlinear kinematic model for the mobile robot, shown in Fig. 3.1, could be represented by

$$\begin{cases} \dot{x}(t) = V(t)\cos\theta(t) \\ \dot{y}(t) = V(t)\sin\theta(t) \\ \dot{\theta}(t) = W(t) \end{cases} \tag{3.1}$$

where V is the linear velocity of the mobile robot, W is its angular velocity, (x, y) is the Cartesian position, and θ is the mobile robot orientation. This model has been widely used to control mobile robots by many authors in the bibliography. An advantage in using this model is that most of the commercial robots have the linear and the angular velocities as the input signals.

Thence, the aim is to find the values of V and W so that the mobile robot follows a pre-established trajectory $(x_{\text{ref}}, y_{\text{ref}})$ with a minimum error. The orientation will be considered as a sacrificed variable to drive the robot position smoothly, following the reference trajectory and its reference, θ_{ref}, will be determined afterwards.

Remark 3.1 The value of the difference between the reference and the real trajectory will be called the tracking error. It is given by $e_x(t) = x_{\text{ref}}(t) - x(t)$ and $e_y(t) = y_{\text{ref}}(t) - y(t)$. The magnitude of the tracking error is given by $\|e_t(t)\| = \sqrt{\left(e_x(t)^2 + e_y(t)^2\right)}$. A similar error can be defined for the orientation, although the reference will be a computed one.

According to the LAB CD methodology described in Chap. 2, a smooth trajectory is sougth for, not only avoiding large tracking errors but also too large control actions. That is, the controlled trajectory is aimed to be:

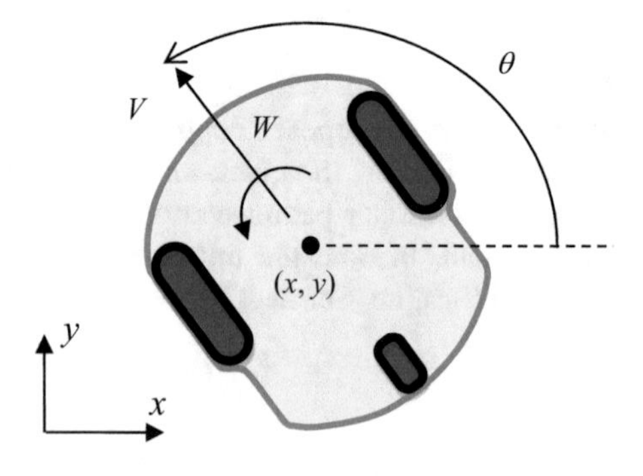

Fig. 3.1 Geometric description of the mobile robot

$$\begin{cases} \dot{x}(t) = \dot{x}_{\text{ref}}(t) + k_x e_x(t) \\ \dot{y}(t) = \dot{y}_{\text{ref}}(t) + k_y e_y(t) \\ \dot{\theta}(t) = \dot{\theta}_{\text{ref}}(t) + k_\theta e_\theta(t) \end{cases} \tag{3.2}$$

where the k coefficients are the controller parameters, assuming a trajectory approaching proportional to the error. To simplify the notation, this expected dynamics can be denoted as

$$\begin{cases} \Delta x = \dot{x}_{\text{ref}}(t) + k_x e_x(t) \\ \Delta y = \dot{y}_{\text{ref}}(t) + k_y e_y(t) \\ \Delta \theta = \dot{\theta}_{\text{ref}}(t) + k_\theta e_\theta(t) \end{cases} \tag{3.3}$$

Thus, combining (3.1)–(3.3), and avoiding the time argument, the control problem is to find the control signals such that the controlled plant behavior is modeled by

$$\begin{bmatrix} \Delta x \\ \Delta y \\ \Delta \theta \end{bmatrix} = \begin{bmatrix} \cos\theta & 0 \\ \sin\theta & 0 \\ 0 & 1 \end{bmatrix} \begin{bmatrix} V \\ W \end{bmatrix} \tag{3.4}$$

which is the equation (2.6) ($b = Au$) for this problem. In order to have an exact solution for this equation, b should be in the column space of A; that is, it should be a linear combination of the columns of A (Strang, 1980). Thus,

$$\begin{aligned} \frac{\Delta x}{\Delta y} &= \frac{V\cos\theta}{V\sin\theta} \\ \Delta\theta &= W \end{aligned} \tag{3.5}$$

To satisfy the first condition, the orientation reference will be chosen such that

$$\theta_{\text{ref}} = \arctan\frac{\Delta y}{\Delta x} \tag{3.6}$$

The control actions are computed solving (3.4) by least squares, once the orientation in the matrix A has been replaced by the required value (3.6) to get an exact solution. That is,

$$\begin{bmatrix} V \\ W \end{bmatrix} = \begin{bmatrix} \Delta x \cos\theta_{\text{ref}} + \Delta y \sin\theta_{\text{ref}} \\ \Delta\theta \end{bmatrix} \tag{3.7}$$

It is worth reminding that the control options are defined in (3.2), thence, the k-parameters are the controller tuning parameters and the controller structure here adopted is based on a proportional to the error approaching.

3.2 Control Performance

In the previous section, the tracking control of the mobile robot modeled by (3.1) to follow a given trajectory has been developed, leading to a feedforward/feedback control law expressed by (3.7). A block diagram of the plant and the controller is depicted in Fig. 3.2, where the different blocks implement the following set of equations

$$\text{Robot (3.1)}: \quad \begin{cases} \dot{x} = V\cos\theta \\ \dot{y} = V\sin\theta \\ \dot{\theta} = W \end{cases}.$$

$$\text{Sacrificed reference (3.6)}: \quad \theta_{\text{ref}} = \arctan\frac{\Delta y}{\Delta x}.$$

$$\text{Controller (3.7)}: \quad \begin{bmatrix} V \\ W \end{bmatrix} = \begin{bmatrix} \Delta x\cos\theta_{\text{ref}} + \Delta y\sin\theta_{\text{ref}} \\ \Delta\theta \end{bmatrix}.$$

$$\text{where (3.3)}: \quad \begin{cases} \Delta x = \dot{x}_{\text{ref}} + k_x e_x \\ \Delta y = \dot{y}_{\text{ref}} + k_y e_y \\ \Delta\theta = \dot{\theta}_{\text{ref}} + k_\theta e_\theta \end{cases} \text{ and } \begin{cases} e_x = x_{\text{ref}} - x \\ e_y = y_{\text{ref}} - y \\ e_\theta = \theta_{\text{ref}} - \theta \end{cases}.$$

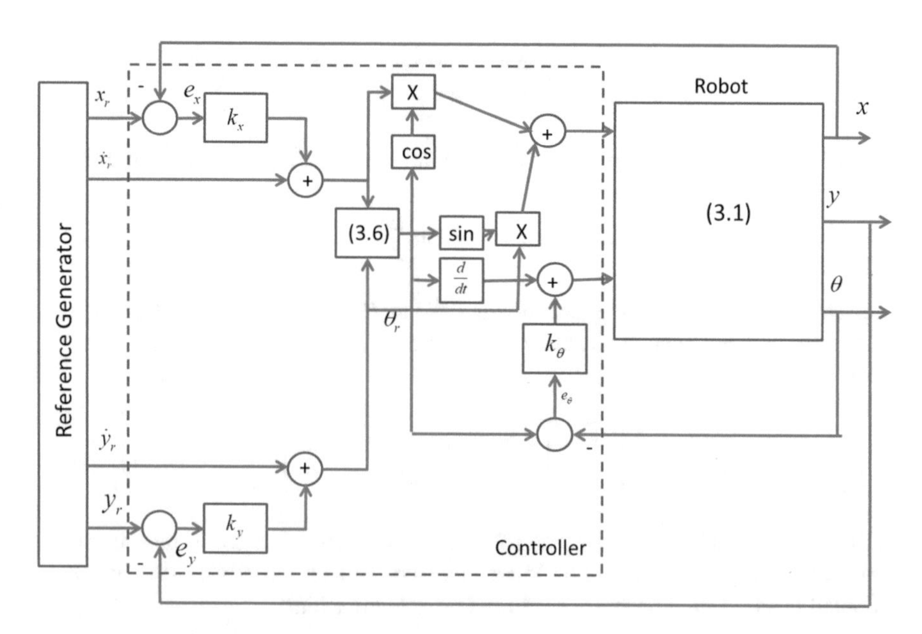

Fig. 3.2 Block diagram of the controlled mobile robot

A Simulink (Matlab®) diagram can be easily implemented to allow the own experimentation by the reader of the proposed control solution. Similar diagrams can be developed for other control schemas proposed in the following chapters.

The performance of the controlled mobile robot is evaluated in steady-state and in the transient behavior. First, the tracking errors are analyzed.

3.2.1 Tracking Errors

The convergence to zero of the tracking errors depends on the controller parameters. The following theorem defines these errors.

Theorem 3.1 If the system behavior is modeled by (3.1) and the controller is designed following (3.3)–(3.7), the tracking errors in following a given trajectory are such that $e(t) \to 0, t \to \infty$ if the controller parameters are positive, that is, $k_x > 0$, $k_y > 0$ and $k_\theta > 0$.

Proof The proof of the convergence to zero of the tracking errors starts with the variable θ. Considering the orientation in the last equation of (3.1) and the angular control action from (3.7), it yields

$$\dot{\theta}(t) = \dot{\theta}_{\text{ref}}(t) + k_\theta(\theta_{\text{ref}}(t) - \theta(t)) \quad \Rightarrow \quad \dot{e}_\theta = -k_\theta e_\theta \tag{3.8}$$

Thus, the dynamics of the orientation error is defined by

$$\dot{\theta}_{\text{ref}}(t) - \dot{\theta}(t) = -k_\theta(\theta_{\text{ref}}(t) - \theta(t)) \Rightarrow e_\theta(t) = e^{-k_\theta t} e_\theta(0) \tag{3.9}$$

and $k_\theta > 0$ ensures the vanishing of this error for $t \to \infty$.

Now, the tracking errors, e_x, e_y, are considered.

Taking into account the Taylor series expression (A.17) for $\cos\theta$

$$\cos\theta = \cos\theta_{\text{ref}} - \sin\underbrace{(\theta_{\text{ref}} + \lambda(\theta_{\text{ref}} - \theta))}_{\theta_\lambda}\underbrace{(\theta - \theta_{\text{ref}})}_{-e_\theta}; 0 < \lambda < 1 \tag{3.10}$$

From the first equation in (3.1) and (3.7)

$$\dot{x}(t) = V\cos\theta = (\Delta x \cos\theta_{\text{ref}} + \Delta y \sin\theta_{\text{ref}})\cos\theta \tag{3.11}$$

it yields

$$\begin{aligned} \dot{x} = (\dot{x}_{\text{ref}} + k_x(x_{\text{ref}} - x))\cos^2\theta_{\text{ref}} + \left(\dot{y}_{\text{ref}} + k_y(y_{\text{ref}} - y)\right)\sin\theta_{\text{ref}}\cos\theta_{\text{ref}} \\ + \underbrace{V\sin\theta_\lambda e_\theta}_{f_\lambda} \end{aligned} \tag{3.12}$$

But, according to (3.6),

$$\left(\dot{y}_{\text{ref}} + k_y e_y\right) \cos \theta_{\text{ref}} = \left(\dot{x}_{\text{ref}} + k_x e_x\right) \sin \theta_{\text{ref}} \tag{3.13}$$

Thence,

$$\dot{x} = \dot{x}_{\text{ref}} + k_x(x_{\text{ref}} - x) + \underbrace{V \sin \theta_\lambda}_{f_\lambda} e_\theta \tag{3.14}$$

$$\dot{e}_x = -k_x e_x - f_\lambda e_\theta \tag{3.15}$$

where

$$f_\lambda = V \sin \theta_\lambda = \left(\Delta x \cos \theta_{\text{ref}} + \Delta y \sin \theta_{\text{ref}}\right) \sin \theta_\lambda \tag{3.16}$$

$$f_\lambda = \left(k_x e_x \cos \theta_{\text{ref}} + k_y e_y \sin \theta_{\text{ref}} + \dot{x}_{\text{ref}} \cos \theta_{\text{ref}} + \dot{y}_{\text{ref}} \sin \theta_{\text{ref}}\right) \sin \theta_\lambda \tag{3.17}$$

In a similar way, for the y variable

$$\dot{y} = V \sin \theta \tag{3.18}$$

By using the Taylor interpolation rule, the function $\sin \theta$ can be expressed as

$$\sin \theta = \sin \theta_{\text{ref}} + \cos \underbrace{\left(\theta_{\text{ref}} + \psi(\theta_{\text{ref}} - \theta)\right)}_{\theta_\psi} \underbrace{(\theta - \theta_{\text{ref}})}_{-e_\theta}; 0 < \psi < 1 \tag{3.19}$$

And again, by using the first row of (3.7) and (3.3) together with (3.19) in (3.18), it yields

$$\dot{y} = \left(\dot{x}_{\text{ref}} + k_x e_x\right) \cos \theta_{\text{ref}} \sin \theta_{\text{ref}} + \left(\dot{y}_{\text{ref}} + k_y e_y\right) \sin^2 \theta_{\text{ref}} + f_\psi e_\theta \tag{3.20}$$

Considering (3.13),

$$\dot{y} = \left(\dot{y}_{\text{ref}} + k_y e_y\right) + f_\psi e_\theta \Rightarrow \dot{e}_y = -k_y e_y - f_\psi e_\theta \tag{3.21}$$

where

$$f_\psi = -\left(k_x e_x \cos \theta_{\text{ref}} + k_y e_y \sin \theta_{\text{ref}} + \dot{x}_{\text{ref}} \cos \theta_{\text{ref}} + \dot{y}_{\text{ref}} \sin \theta_{\text{ref}}\right) \cos \theta_\psi \tag{3.22}$$

Considering altogether (3.8) and (3.17) and (3.21) in matrix form, it yields

$$\begin{bmatrix} \dot{e}_x \\ \dot{e}_y \end{bmatrix} = \begin{bmatrix} -k_x & 0 \\ 0 & -k_y \end{bmatrix} \begin{bmatrix} e_x \\ e_y \end{bmatrix} + \underbrace{\begin{bmatrix} -f_\lambda \\ -f_\psi \end{bmatrix} e_\theta}_{\substack{\text{Bounded} \\ \text{nonlinearity} \to 0}} \tag{3.23a}$$

$$\dot{e}_\theta = -k_\theta e_\theta \tag{3.23b}$$

Thus, for $0 < k_\theta$, the error $e_\theta(t) \to 0$, $t \to \infty$ in (3.23b) tends asymptotically to zero and for $0 < k_x$, $0 < k_y$ the tracking errors in (3.23a) also tend to zero as the nonlinearity in their dynamic behavior is bounded and tends to zero. That is, the tracking error

$$e = \left[e_x e_y e_\theta\right]^T; \lim_{t \to \infty} e(t) = 0 \tag{3.24}$$

tends to zero.

Remark 3.2 The equations (3.23a) and (3.23b) represent the dynamics of a linear system with an added nonlinearity that tends to zero. It can be shown that the nonlinearity is bounded in the same manner as it was shown for other functions in Scaglia et al. (2019). It also implies that, in general, the tracking error in the sacrificed variable is brought to zero, and later on all the tracking errors disappear.

3.2.2 Controlled Plant Stability

The controlled robot is modeled by the set of equations summarized at the beginning of this section. In order to prove the stability of the controlled plant, the following theorem is stated.

Theorem 3.2 If the system behavior is modeled by (3.1), and the controller is designed following (3.3)–(3.7), the controlled plant in following a given trajectory is stable if the controller parameters are positive, that is, $k_x > 0$, $k_y > 0$ and $k_\theta > 0$.

Proof The system (3.23a) and (3.23b) can be considered as a cascade system, where the second part (3.23b) is stable if $k_\theta > 0$. On the other hand, the first part is globally uniformly asymptotically stable at the origin if $k_x > 0$, $k_y > 0$. That is, if (3.23a) is stable for $e_\theta(t) = 0$. Thus, by the stability Lemma for cascade systems (Lemma 4.7, Khalil & Grizzle, (2002)), the full system is stable.

3.2.3 Experimental Results

In the next chapter, different control solutions for the mobile robot are developed, the results being compared with those reported in the literature. The control structure defined in this chapter is the simplest one, and it provides satisfactory controlled plant behavior.

3.3 Controller Tuning

It has been proven that the controlled robot is stable and asymptotically follows the required trajectory. The only parameters to be tuned are those in the trajectory approaching, and it has been proven that being positive is the only requirement to fulfill the basic performance.

In practical applications, the differences in the controller responses for two different sets of gains occur every time the robot has to turn, for example, a corner of the reference trajectory. However, once the robot is aligned with the trajectory (i.e., its pose is kinematically compatible with the path), the performance of the controller is the same for both set of gains (Cheein, Blazic, & Torres-Torriti, 2015).

Nowadays, if there is a requirement for higher performance, such as faster reply, reduced overshoot, or any other dynamic requirement, there are no analytic formulas to compute the controller parameters.

In the field of systems and control, probabilistic methods have been found useful, especially for problems related to robustness of uncertain systems. One of these methods, the Monte Carlo Randomized Algorithm, is widely used in many fields such as the radioactive decay, systems of interacting atoms, the traffic on roads, etc. In the control area, Monte Carlo methods (MCMs) allow the estimation of an expectation value, and they provide effective tools for the analysis of probabilistically robust control schemes.

Monte Carlo method consists of a random algorithm that makes random choices to produce a result. This implies that it can give a higher or lower index, that is, it is subject to probability. Monte Carlo is a random algorithm that may not find the minimum, but the probability of such a result is limited. One way to reduce this probability is simply to execute the algorithm repeatedly (N times) with random options independent of the parameters at each moment (see Theorem 2 of (Tempo & Ishii, 2007)). The number of simulations (N) is obtained by choosing an appropriate accuracy and confidence to limit the possibility of an unwanted response, then used (3.25) to calculate the number of iterations.

$$N \geq \left[\frac{\log \frac{1}{\delta}}{\log \frac{1}{1-\varepsilon}} \right] \tag{3.25}$$

where δ is confidence and ε is accuracy.

This algorithm is very useful to select a set of controller gains, and it will be used to define the controller parameters.

The sequence to apply the Monte Carlo algorithm is:

1. Define the controller's parameters to be optimized.
2. Determine the number of simulations to be performed (N). The values of δ and ε are chosen, depending on the desired accuracy, for example: $\delta = 0.01$ and $\varepsilon = 0.005$, resulting in $N = 1000$.
3. A value is randomly assigned for each parameter of the controller.
4. The process is simulated and cost index (J) is calculated.
5. Repeat steps 3 and 4 until completing the N iterations.
6. Finally, the controller parameters minimizing J are selected.

The parameters of the LAB controllers developed and used in the simulations and experimentation carried out in this book were selected following the sequence above.

References

Cheein, F. A., Blazic, S., & Torres-Torriti, M. (2015, September). Computational approaches for improving the performance of path tracking controllers for mobile robots. In *2015 IEEE/RSJ International Conference on Intelligent Robots and Systems (IROS)* (pp. 6495–6500). New York: IEEE.

Li, Z., Deng, J., Lu, R., Xu, Y., Bai, J., & Su, C. Y. (2015). Trajectory-tracking control of mobile robot systems incorporating neural-dynamic optimized model predictive approach. *IEEE Transactions on Systems, Man, and Cybernetics: Systems, 46*(6), 740–749.

Khalil, H. K., & Grizzle, J. W. (2002). *Nonlinear systems (Vol. 3)*. (p 180), Upper Saddle River, NJ: Prentice hall

Panahandeh, P., Alipour, K., Tarvirdizadeh, B., & Hadi, A. (2019). A self-tuning trajectory tracking controller for wheeled mobile robots. *Industrial Robot: The International Journal of Robotics Research and Application, 46*(6), 828–838.

Proaño, P., Capito, L., Rosales, A., & Camacho, O. (2015, July). Sliding mode control: Implementation like PID for trajectory-tracking for mobile robots. In *2015 Asia-Pacific Conference on Computer Aided System Engineering* (pp. 220–225). New York: IEEE.

Scaglia, G., Montoya, L. Q., Mut, V., & Di Sciascio, F. (2009). Numerical methods based controller design for mobile robots. *Robotica, 27*(2), 269–279.

Scaglia, G., Serrano, E., Rosales, A., & Albertos, P. (2019). Tracking control design in nonlinear multivariable systems: Robotic applications. *Mathematical Problems in Engineering, 2019*, 8643515. https://doi.org/10.1155/2019/8643515.

Scaglia, G., Quintero, O. L., Mut, V., & di Sciascio, F. (2008). Numerical methods based controller design for mobile robots. *In IFAC Proceedings Volumes, 41*(2), 4820–4827. https://doi.org/10.3182/20080706-5-KR-1001.00810.

Serrano, M. E., Godoy, S. A., Quintero, L., & Scaglia, G. J. (2017). Interpolation based controller for trajectory tracking in mobile robots. *Journal of Intelligent and Robotic Systems, 86*(3–4), 569–581.

Strang, G. (1980). *Linear algebra and its applications*. New York: Academic Press.

Sun, W., Tang, S., Gao, H., & Zhao, J. (2016). Two time-scale tracking control of nonholonomic wheeled mobile robots. *IEEE Transactions on Control Systems Technology, 24*(6), 2059–2069.

Tempo, R., & Ishii, H. (2007). Monte Carlo and Las Vegas randomized algorithms for systems and control: An introduction. *European Journal of Control, 13*(2–3), 189–203. https://doi.org/10.3166/ejc.13.189-203.

Chapter 4
Discrete Time Control of a Mobile Robot

In the previous chapter, the Linear Algebra-Based Control Design methodology has been applied to design the continuous time control of a simple kinematic model of a mobile robot. But, nowadays, all the controllers are implemented in digital systems. Thus, the controllers should be designed in discrete time or the designed CT controller should be discretized before being implemented. One of the main features of a CD methodology is its applicability to simple and approximate plant models, to avoid difficult to design and implement complex controllers.

In this chapter, this control design methodology will be applied in DT and several models and discretization approaches for the same process will be considered. The simpler the model is, the easier the controller results, and its tuning also becomes easier. The advantages and drawbacks of designing the control by using more complicated models will be outlined and the results compared in a practical experiment. This comparison will include the CT controller developed in the previous chapter. The chapter is structured as follows: first, the kinematic model (3.1) will be discretized, following different approximations. Then, the dynamic model will be used. A practical experiment will confirm the expected results, and some conclusions will be drafted (Scaglia, Serrano, Rosales, & Albertos, 2019).

4.1 Kinematic Discrete Time Control

A nonlinear kinematic model for a mobile robot was described in (3.1), repeated here for completeness

G. Scaglia et al., *Linear Algebra Based Controllers*,
https://doi.org/10.1007/978-3-030-42818-1_4

$$\begin{cases} \dot{x} = V \cos \theta \\ \dot{y} = V \sin \theta \\ \dot{\theta} = W \end{cases} \tag{4.1}$$

The goal is to find the values of V and W so that the mobile robot follows a pre-established trajectory $(x_{\text{ref}}(t), y_{\text{ref}}(t))$ with a minimum error. Now, the values of $x(t)$, $y(t)$, $\theta(t)$, $V(t)$, and $W(t)$ are considered at discrete time $t = nT$, where T is the sampling period and $n \in \{0, 1, 2, \cdots\}$, being denoted as x_n, y_n, θ_n, V_n, and W_n, respectively.

4.1.1 Control Design

In order to apply the LAB CD methodology described in Chap. 3 acting on DT, the robot model given by (4.1) is discretized. Initially, the first-order Euler's approximation is chosen (Scaglia et al., 2019; Scaglia, Quintero, Mut, & di Sciascio, 2008; Serrano, Godoy, Mut, Ortiz, & Scaglia, 2016; Serrano, Scaglia, Cheein, Mut, & Ortiz, 2015):

$$\begin{cases} x_{n+1} = x_n + T \ V_n \ \cos \theta_n \\ y_{n+1} = y_n + T \ V_n \ \sin \theta_n \\ \theta_{n+1} = \theta_n + T \ W_n \end{cases} \tag{4.2}$$

Other discretization approaches could be considered, as described later in the chapter.

Next, (4.2) is rearranged in matrix form as

$$\begin{bmatrix} x_{n+1} \\ y_{n+1} \\ \theta_{n+1} \end{bmatrix} = \begin{bmatrix} x_n \\ y_n \\ \theta_n \end{bmatrix} + T \begin{bmatrix} \cos \theta_n & 0 \\ \sin \theta_n & 0 \\ 0 & 1 \end{bmatrix} \begin{bmatrix} V_n \\ W_n \end{bmatrix} \tag{4.3}$$

and the position value at the next time instant will be replaced by the desired value, assuming a proportional approaching. Again, more elaborated approaching strategies could be considered.

The necessary orientation to make the mobile robot tend to the reference trajectory will be denoted as the reference for the sacrificed variable (θ_{ref}). Similar to the x and y variables, the orientation value at the next instant will be replaced by $\theta_{\text{ref},n+1}$, assuming a proportional approaching. The positive constants k_x, k_y, and k_θ allow the tuning of the performance of the proposed controlled system

$$\begin{bmatrix} \cos\theta_n & 0 \\ \sin\theta_n & 0 \\ 0 & 1 \end{bmatrix} \begin{bmatrix} V_n \\ W_n \end{bmatrix} = \frac{1}{T} \begin{bmatrix} x_{\text{ref},n+1} - k_x(x_{\text{ref},n} - x_n) - x_n \\ y_{\text{ref},n+1} - k_y(y_{\text{ref},n} - y_n) - y_n \\ \theta_{\text{ref},n+1} - k_\theta(\theta_{\text{ref},n} - \theta_n) - \theta_n \end{bmatrix} \tag{4.4}$$

The value of $\theta_{\text{ref},n}$ is calculated to fulfill the conditions in order that system (4.4) has an exact solution, thence:

$$\tan\theta_{\text{ref},n} = \frac{\sin\theta_{\text{ref},n}}{\cos\theta_{\text{ref},n}} = \frac{\Delta y}{\Delta x} = \frac{y_{\text{ref},n+1} - k_x(y_{\text{ref},n} - y_n) - y_n}{x_{\text{ref},n+1} - k_y(x_{\text{ref},n} - x_n) - x_n} \tag{4.5}$$

The orientation θ_{ref} represents the required orientation to make the mobile robot tend to the reference trajectory.

Finally, the proposed control law is obtained, solving the system (4.4)

$$\begin{bmatrix} V_n \\ W_n \end{bmatrix} = \begin{bmatrix} \frac{x_{\text{ref},n+1} - k_x(x_{\text{ref},n} - x_n) - x_n}{T} \cos\theta_{\text{ref},n} + \frac{y_{\text{ref},n+1} - k_y(y_{\text{ref},n} - y_n) - y_n}{T} \sin\theta_{\text{ref},n} \\ \frac{\theta_{\text{ref},n+1} - k_\theta(\theta_{\text{ref},n} - \theta_n) - \theta_n}{T} \end{bmatrix} \tag{4.6}$$

It is worth noting that in (4.5), the computed orientation is $\theta_{\text{ref},n}$, but $\theta_{\text{ref},n+1}$ is required to evaluate the control action in (4.6). That means that this value should be estimated from the previous knowledge of the series $\{\theta_{\text{ref},\, i}\}$.

4.1.2 Control Performance

In order to analyze the controlled plant behavior, the tracking errors are evaluated.

Theorem 4.1 If the system behavior is modeled by (4.2) and the controller is designed by (4.6), the tracking errors in following a given trajectory are such that $e_n \to 0,\ n \to \infty$ if the controller parameters are $|k_x| < 1$, $|k_y| < 1$ and $|k_\theta| < 1$. Moreover, the controlled system is stable.

Proof The proof of the convergence to zero of the tracking errors as well as the stability is similar to that developed in Theorem 3.1, and it will be omitted here. Also, the reasoning in Theorem 3.2 leads to prove the stability of this controller. (A detailed development can be found in Scaglia et al., 2010).

To avoid undesirable oscillations, the controller parameters should be positive.

4.1.3 Experimental Results

The practical implementation of the controller described earlier is shown in this section, using a PIONEER 3AT mobile robot. Although the PIONEER 3AT mobile robot includes an estimation system based on odometric positioning system, an updating through external sensors is necessary. But this problem is a different one from the strategy of trajectory tracking, and it is not considered in this book.

The mobile robot and the laboratory facilities at the Instituto de Automática, Facultad de Ingeniería (INAUT), Universidad Nacional de San Juan, where the experiments were carried out, are shown in Fig. 4.1. The structure of the control system is shown in Fig. 4.2.

The experimentation was carried out, assuming that the robot should follow an eight-shaped trajectory. The initial condition for the robot mobile position is the system origin, and the trajectory begins at the position $(x_{\text{ref}}(0), y_{\text{ref}}(0)) = (1\text{ m}, -1\text{ m})$, whereas the used sampling period is $T = 0.1$ s. The adopted controller parameter values are $k_x = k_y = 0.94$ and $k_\theta = 0.92$. These values are selected to ensure that the tracking error tends to zero, and that the robot has a smooth tracking.

The results of the controller implementation are shown in Fig. 4.3. As it can be seen, the robot reaches and quickly follows the desired trajectory. In addition, in Fig. 4.4, it is shown that the tracking error remains close to zero once the robot

Fig. 4.1 The PIONEER 3AT and the INAUT laboratory

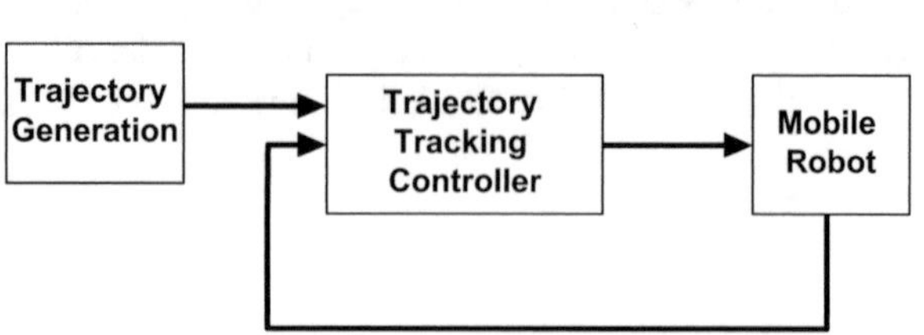

Fig. 4.2 General structure of the proposed controller

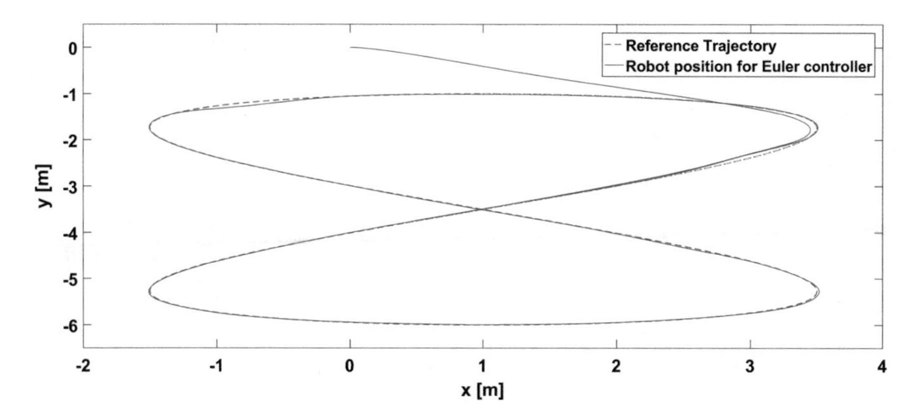

Fig. 4.3 Reference trajectory and robot position

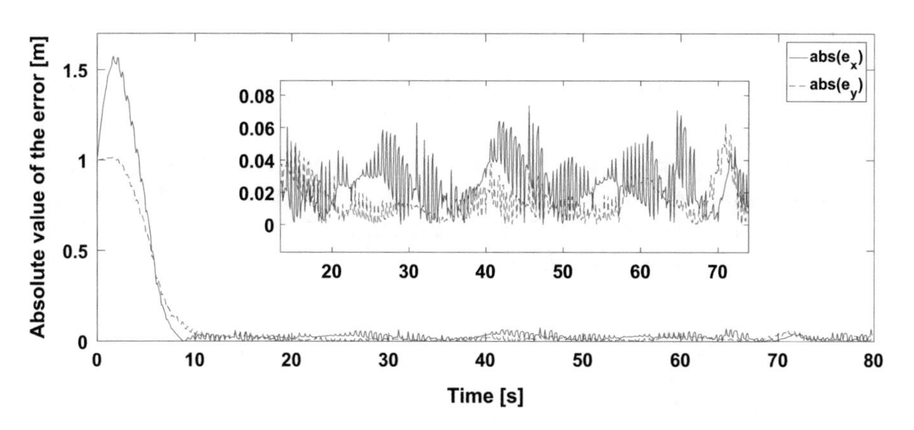

Fig. 4.4 Absolute values of the tracking error in x-coordinate and y-coordinate

reaches the reference. The control actions calculated by the controller (4.6) and the robot velocities are shown in Fig. 4.5.

4.1.4 Discrete Time Model of the Robot with Trapezoidal Approximation

The LAB procedure presented earlier is now applied to the robot model by using a trapezoidal approximation of the kinematic robot model. This is a second-order approximation of the derivative, which is replaced by, for instance,

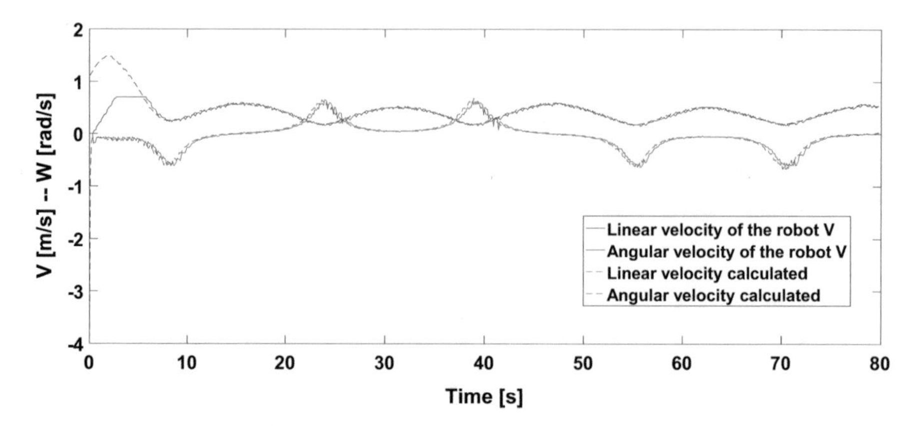

Fig. 4.5 Calculated control actions and robot velocities

$$\dot{x} = V\cos\theta \Rightarrow \quad x_{n+1} = x_n + \frac{T}{2}(V_{n+1}\cos\theta_{n+1} + V_n\cos\theta_n) \tag{4.7}$$

This approach implies that the robot velocities will appear in the control law, and the achieved performance of the system could be improved. If the DT trapezoidal approximation is applied to the CT kinematic model of the mobile robot (3.1), the following set of equations is obtained:

$$\begin{bmatrix} x_{n+1} \\ y_{n+1} \\ \theta_{n+1} \end{bmatrix} = \begin{bmatrix} x_n \\ y_n \\ \theta_n \end{bmatrix} + \frac{T}{2} \times \left(\begin{bmatrix} \cos\theta_{n+1} & 0 \\ \sin\theta_{n+1} & 0 \\ 0 & 1 \end{bmatrix} \begin{bmatrix} V_{n+1} \\ W_{n+1} \end{bmatrix} + \begin{bmatrix} \cos\theta_n & 0 \\ \sin\theta_n & 0 \\ 0 & 1 \end{bmatrix} \begin{bmatrix} V_n \\ W_n \end{bmatrix} \right) \tag{4.8}$$

Leading to the inverse model

$$\begin{bmatrix} \cos\theta_{n+1} & 0 \\ \sin\theta_{n+1} & 0 \\ 0 & 1 \end{bmatrix} \begin{bmatrix} V_{n+1} \\ W_{n+1} \end{bmatrix} = \begin{bmatrix} \frac{2}{T}(x_{n+1} - x_n) - V_n\cos\theta_n \\ \frac{2}{T}(y_{n+1} - y_n) - V_n\sin\theta_n \\ \frac{2}{T}(\theta_{n+1} - \theta_n) - W_n \end{bmatrix} \tag{4.9}$$

Next, the reference position of the mobile robot, the reference velocities, and the required orientation of the mobile robot are replaced in (4.9), assuming a proportional approaching. Again, the positive constants k_x, k_y, and k_θ are the controller parameters.

$$\begin{bmatrix} \cos\theta_{n+1} & 0 \\ \sin\theta_{n+1} & 0 \\ 0 & 1 \end{bmatrix} \begin{bmatrix} V_{n+1} \\ W_{n+1} \end{bmatrix} = \begin{bmatrix} \frac{2\Delta x}{T} - V_n \cos\theta_n \\ \frac{2\Delta y}{T} - V_n \sin\theta_n \\ \frac{2\Delta\theta}{T} - W_n \end{bmatrix} \tag{4.10}$$

where

$$\begin{aligned} \Delta x &= x_{\text{ref},n+1} - k_x(x_{\text{ref},n} - x_n) - x_n \\ \Delta y &= y_{\text{ref},n+1} - k_y(y_{\text{ref},n} - y_n) - y_n \\ \Delta\theta &= \theta_{\text{ref},n+1} - k_\theta(\theta_{\text{ref},n} - \theta_n) - \theta_n \end{aligned} \tag{4.11}$$

The value of $\theta_{\text{ref},n+1}$ is obtained by searching the conditions such that (4.10) has an exact solution:

$$\tan\theta_{\text{ref},n+1} = \frac{\frac{2}{T}\left(y_{\text{ref},n+1} - k_y\left(y_{\text{ref},n} - y_n\right) - y_n\right) - V_n \sin\theta_n}{\frac{2}{T}\left(x_{\text{ref},n+1} - k_x(x_{\text{ref},n} - x_n) - x_n\right) - V_n \cos\theta_n} \tag{4.12}$$

and the proposed controller is obtained by solving the system (4.9) using least square optimization

$$\begin{aligned} V_{n+1} &= \left(\frac{2}{T}\Delta x - V_n \cos\theta_n\right) \cos\theta_{\text{ref},n+1} + \left(\frac{2}{T_s}\Delta y - V_n \sin\theta_n\right) \sin\theta_{\text{ref},n+1} \\ W_{n+1} &= \frac{2}{T}\Delta\theta - W_n \end{aligned} \tag{4.13}$$

4.1.5 *Performance of the Trapezoidal Controller*

It would be interesting to compare the performance of the controller obtained through the trapezoidal DT approximation with that obtained using the Euler's one to see the possible advantages obtained by using a slightly more complicated model. In what follows, the tracking errors and dynamics of the robot controlled by the trapezoidal controller are discussed.

Theorem 4.2 If the system behavior is modeled by (4.8) and the controller is designed by (4.13), the tracking errors in following a given trajectory are such that $e_n \to 0$, $n \to \infty$ if the controller parameters are $|k_x| < 1$, $|k_y| < 1$ and $|k_\theta| < 1$. Moreover, the controlled system is stable.

The proof of convergence to zero of the tracking errors is similar to the one developed in the previous case. Considering the orientation from the third row in (4.8) and the control action from (4.13)

$$\theta_{n+1} = \theta_n + \frac{T}{2}(W_n + W_{n+1}) \tag{4.14}$$

$$W_{n+1} = \frac{2}{T}(\theta_{\mathrm{ref},n+1} - k_\theta(\theta_{\mathrm{ref},n} - \theta_n) - \theta_n) - W_n \tag{4.15}$$

By replacing the control action W_{n+1} given by (4.15) in (4.14), the following expression is found

$$\theta_{n+1} = \theta_n + \frac{T}{2}\left(W_n + \frac{2}{T}(\theta_{\mathrm{ref},n+1} - k_\theta(\theta_{\mathrm{ref},n} - \theta_n) - \theta_n) - W_n\right) \tag{4.16}$$

After some simple operations, it yields:

$$\theta_{\mathrm{ref},n+1} - \theta_{n+1} - k_\theta(\theta_{\mathrm{ref},n} - \theta_n) = 0 \tag{4.17}$$

Thence, the error dynamics for the orientation can be expressed as

$$e_{\theta,n+1} - k_\theta e_{\theta,n} = 0 \tag{4.18}$$

Thus, the error tends asymptotically to zero if

$$|k_\theta| < 1 \Rightarrow -1 < k_\theta < 1 \tag{4.19}$$

Moreover, if $0 < k_\theta < 1$ and $n \rightarrow \infty$ (with $n \in N$), then $e_{\theta,\ n\ +\ 1} \rightarrow 0$ without oscillations.

Now, the convergence analysis of e_x and e_y is developed below following the same reasoning than before. From the first row of the system model (4.8)

$$x_{n+1} = x_n + \frac{T}{2}(V_n \cos\theta_n + V_{n+1}\cos\theta_{n+1})$$

and by using the Taylor interpolation rule in a similar way to (3.10), following the development (3.11) to (3.17), it can be shown that

$$x_{n+1} = x_n + \frac{T}{2}\left(V_n \cos\theta_n + \left(\frac{2}{T}\Delta x - V_n\cos\theta_n\right) + \underbrace{V_{n+1}\sin\theta_{\lambda,n} e_{\theta,n+1}}_{f_{\lambda,n}}\right) \tag{4.20}$$

According to (4.20),

$$x_{\text{ref},n+1} - x_{n+1} - k_x(x_{\text{ref},n} - x_n) + \frac{T}{2} f_{\lambda,n} e_{\theta,n+1} = 0 \tag{4.21}$$

Thence,

$$e_{x,n+1} = k_x e_{x,n} - \frac{T}{2} f_{\lambda,n} e_{\theta,n+1} \tag{4.22}$$

Now, the same reasoning is applied to the y-coordinate, by starting with the second row in the system model (4.8)

$$y_{n+1} = y_n + \frac{T}{2}(V_n \sin\theta_n + V_{n+1} \sin\theta_{n+1}) \tag{4.23}$$

By using the Taylor interpolation rule for the sine function (3.19), (4.23) can be expressed as

$$y_{n+1} = y_n + \frac{T}{2}\left(V_n \sin\theta_n + V_{n+1} \sin\theta_{\text{ref},n+1} - \underbrace{V_{n+1} \cos\theta_{\psi,n}}_{f_{\psi,n}} e_{\theta,n+1}\right) \tag{4.24}$$

Then, considering the control action V_{n+1} (4.13) and multiplying by $\sin\theta_{\text{ref},\, n+1}$, after some simple operations

$$V_{n+1} \sin\theta_{\text{ref},n+1} = \frac{2}{T}\Delta y - V_n \sin\theta_n \tag{4.25}$$

Thus,

$$y_{n+1} = y_n + \frac{T_s}{2}\left(V_n \sin\theta_n + \left(\frac{2}{T_s}\Delta y - V_n \sin\theta_n\right) - f_{\psi,n} e_{\theta,n+1}\right) \tag{4.26}$$

and, taking into account (4.11),

$$y_{\text{ref},n+1} - y_{n+1} - k_y\left(y_{\text{ref},n} - y_n\right) - \frac{T}{2} f_{\psi,n} e_{\theta,n+1} = 0 \tag{4.27}$$

leading to

$$e_{y,n+1} - k_y e_{y,n} - T f_{\psi,n} e_{\theta,n+1} = 0 \tag{4.28}$$

Considering (4.18), (4.22), and (4.28) it yields, in compact form

$$\begin{bmatrix} e_{x,n+1} \\ e_{y,n+1} \\ e_{\theta,n+1} \end{bmatrix} = \begin{bmatrix} k_x & 0 & -\frac{T}{2} f_{\lambda,n} \\ 0 & k_y & \frac{T}{2} f_{\psi,n} \\ 0 & 0 & k_\theta \end{bmatrix} \begin{bmatrix} e_{x,n} \\ e_{y,n} \\ e_{\theta,n} \end{bmatrix} \tag{4.29}$$

Thus the orientation error $e_{\theta,\ n}$ tends to zero as far as $0 < |k_\theta| < 1$, and the trajectory error tends to zero if the controller coefficients are also in the same range, as the nonlinear terms multiplying the orientation error are bounded.

4.1.6 Experimental Results

To illustrate the controller performance, the test performed in the previous section is repeated here but now using the trapezoidal controller developed earlier (4.13). The conditions to run the test are the same as in the previous section. Moreover, the reference trajectory and the initial conditions of the robot are also the same. The used sampling period is $T = 0.1$ s. The controller parameter values are $k_x = k_y = 0.94$ and $k_\theta = 0.92$, being selected to ensure that the tracking error tends to zero, and that the robot has a smooth tracking.

As a result of the test, in Fig. 4.6, it is shown that the robot can follow the proposed reference trajectory without undesired oscillations. By inspection of Fig. 4.7, it can be seen that the error remains close to zero. The robot velocities and the control action calculated at each sample time using equation (4.13) are shown in Fig. 4.8.

Comparing Figs. 4.3, 4.4, 4.6, and 4.7, a small reduction of the errors can be observed, and from Figs. 4.5 and 4.8, it can be seen that the control actions are similar, although a larger control action is initially needed by using the trapezoidal model.

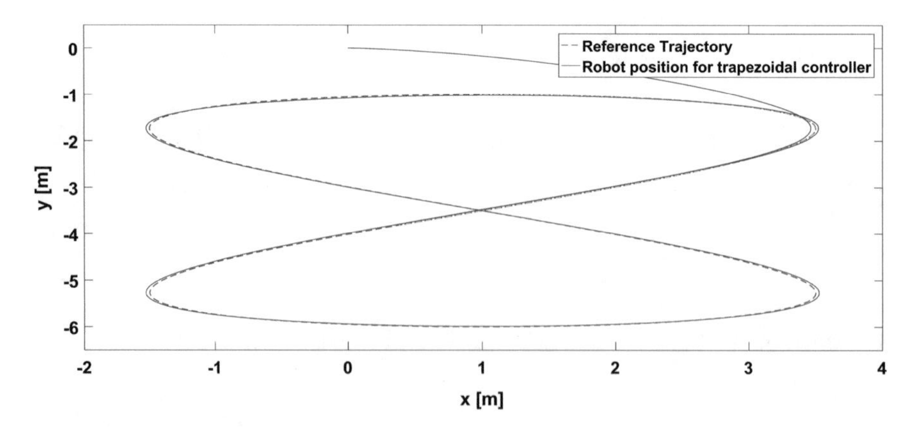

Fig. 4.6 Reference trajectory and robot position

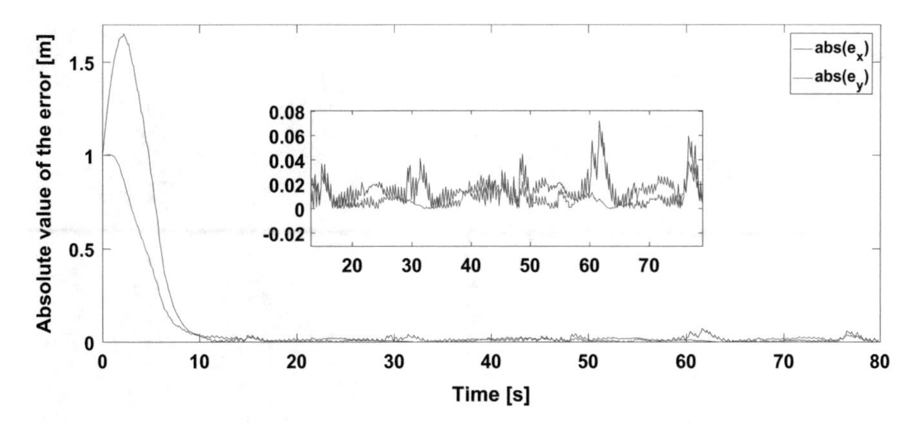

Fig. 4.7 Absolute values of the tracking error in x-coordinate and y-coordinate

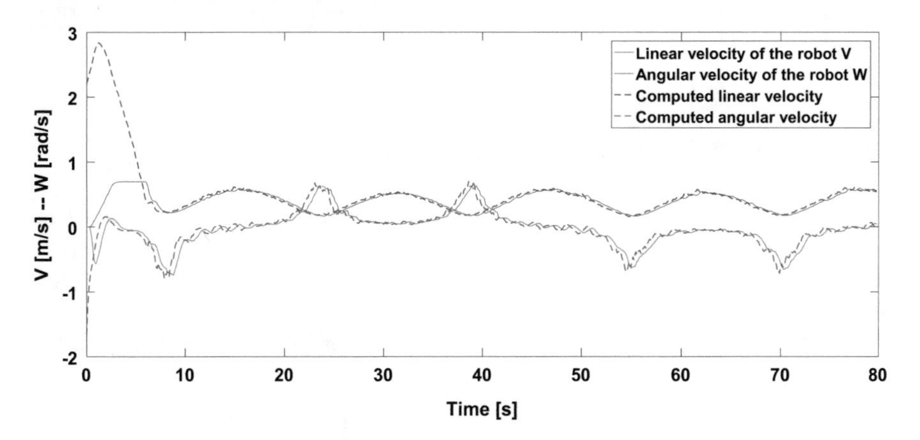

Fig. 4.8 Computed control actions and robot velocities

4.2 Dynamic Model of the Robot

To perform tasks requiring high speed and/or heavy loads transportation, it is very important to consider the dynamics of the mobile robot because such tasks exert large external forces on the robot and they will unavoidably influence its path and direction. Thence, a kinematic model is not sufficient. Moreover, dynamic characteristics of the robot such as mass and inertia center change according to the robot load. A non-holonomic dynamic model of a unicycle-like mobile robot (De La Cruz & Carelli, 2008) is shown in Fig. 4.9a, and it is represented by (4.30).

The robot position is defined by (x, y); this point is located at a distance a from the rear axis center of the robot, u and $\bar{u}$ are the longitudinal and side speeds of the mass center, ω is the angular speed and ψ is the orientation angle, G is the gravity center, and B is the baseline center of the wheels; $F_{rrx'}$ and $F_{rry'}$ are the longitudinal and lateral tire forces of the right wheel; $F_{rlx'}$ and $F_{rly'}$ are the longitudinal and lateral tire

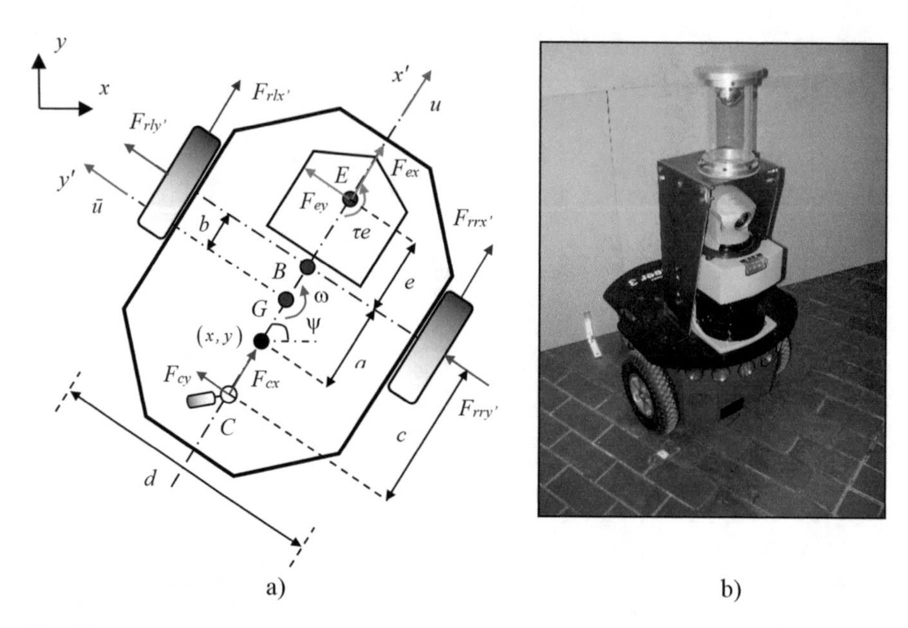

Fig. 4.9 PIONEER 3DX mobile robot, (**a**) model; (**b**) laboratory equipment

forces of the left wheel; $F_{cx'}$ and $F_{cy'}$ are the longitudinal and lateral forces exerted on C by the castor; $F_{ex'}$ and $F_{ey'}$ are the longitudinal and lateral forces exerted on E by the tool (e.g., a robotic arm); b, c, d, and e are distances; and τ_e is the moment exerted by the tool.

From the diagram in Fig. 4.9a, the dynamic model of the mobile robot is given by:

$$\begin{bmatrix} \dot{x} \\ \dot{y} \\ \dot{\psi} \\ \dot{u} \\ \dot{\omega} \end{bmatrix} = \begin{bmatrix} u\cos\psi - a\omega\sin\psi \\ u\sin\psi + a\omega\cos\psi \\ \omega \\ \frac{\theta_3}{\theta_1}\omega^2 - \frac{\theta_4}{\theta_1}u \\ -\frac{\theta_5}{\theta_2}u\omega - \frac{\theta_6}{\theta_2}\omega \end{bmatrix} + \begin{bmatrix} 0 & 0 \\ 0 & 0 \\ 0 & 0 \\ \frac{1}{\theta_1} & 0 \\ 0 & \frac{1}{\theta_2} \end{bmatrix} \begin{bmatrix} uc \\ \omega c \end{bmatrix} + \begin{bmatrix} \delta_x \\ \delta_y \\ 0 \\ \delta_u \\ \delta_\omega \end{bmatrix} \tag{4.30}$$

The identified parameters $\boldsymbol{\theta}$ of the dynamic model (validated in De La Cruz & Carelli, 2008) are:

$$\begin{aligned} \theta_1 &= \left(\frac{R_a}{k_a}(mR_tr + 2I_e) + 2rk_{DT}\right)/(2rk_{PT}) = 0.24089 & \theta_4 &= \frac{R_a}{k_a}\left(\frac{k_ak_b}{R_a} + B_e\right)/(rk_{PT}) + 1 = 0.99629 \\ \theta_2 &= \left(\frac{R_a}{k_a}(I_ed^2 + 2R_tr(I_z + mb^2)) + 2rdk_{DR}\right)/(2rdk_{PR}) = 0.2424 & \theta_5 &= \frac{R_a}{k_a}mbR_t/(dk_{PR}) = -0.0037256 \\ \theta_3 &= \frac{R_a}{k_a}mbR_t/(2k_{PT}) = -0.00093603 & \theta_6 &= \frac{R_a}{k_a}\left(\frac{k_ak_b}{R_a} + B_e\right)d/(2rk_{PR}) + 1 = 1.0915 \end{aligned} \tag{4.31}$$

where m is the robot mass; r is the right and left wheel radius; k_b is equal to the voltage constant multiplied by the gear ratio; R_a is the electric resistance constant; k_a is the torque constant multiplied by the gear ratio; k_{PR}, k_{PT}, and k_{DT} are positive constants; I_e and B_e are, respectively, the moment of inertia and the viscous friction coefficient of the combined motor rotor, gearbox, and wheel; and R_t is the nominal radius of the tire.

The elements of the uncertainty vector $\boldsymbol{\delta}$, related to the mobile robot, are $\boldsymbol{\delta} = \left[\delta_x\ \delta_y\ 0\ \delta_u\ \delta_\omega\right]^T$, where δ_x and δ_y depend on velocities due to wheels slide and robot orientation and δ_u and δ_ω depend on mechanic parameters of the robot, such as mass, inertial moment, wheel diameter, engine and servo-controllers parameters, forces on the wheels, etc. All these parameters are considered as disturbances.

Remark 4.3 If the slip speeds of the wheels, the forces and torques exerted by the tool, and the forces exerted by the castor wheel are of no significant value, the uncertainties vector $\boldsymbol{\delta}$ will not be considered.

In general, most market-available robots have low-level PID velocity controllers to track input reference velocities, not allowing the motor voltage to be directly driven. Therefore, it is useful to express the mobile robot model in a suitable way by considering rotational and translational reference velocities as control signals. We assume that the mobile robot is moving on a horizontal plane without slip.

4.2.1 *Linear Algebra-Based Controller*

The control objective is to follow a reference trajectory (x_{ref}, y_{ref}) by operating the velocities u_c and ω_c. To apply the LAB controller design method in DT, a discrete approximation of the dynamic model of mobile robot (4.32) is first obtained (Rosales, Scaglia, Mut, & di Sciascio, 2009),

$$\begin{bmatrix} x_{n+1} \\ y_{n+1} \\ \psi_{n+1} \\ u_{n+1} \\ \omega_{n+1} \end{bmatrix} = \begin{bmatrix} x_n \\ y_n \\ \psi_n \\ u_n \\ \omega_n \end{bmatrix} + T \left\{ \begin{bmatrix} u_n \cos\psi_n \\ u_n \sin\psi_n \\ \omega_n \\ \frac{\theta_3}{\theta_1}\omega_n^2 - \frac{\theta_4}{\theta_1}u_n \\ -\frac{\theta_5}{\theta_2}u_n\omega_n - \frac{\theta_6}{\theta_2}\omega_n \end{bmatrix} + \begin{bmatrix} 0 & 0 \\ 0 & 0 \\ 0 & 0 \\ \frac{1}{\theta_1} & 0 \\ 0 & \frac{1}{\theta_2} \end{bmatrix} \begin{bmatrix} u_{c,n} \\ \omega_{c,n} \end{bmatrix} \right\} \tag{4.32}$$

In this case, there are three sacrificed variables, ψ_n, u_n and ω_n. Then, the values of the reference trajectory and that of the sacrificed values are replaced in system (4.32), considering a proportional approach to reduce the tracking error

$$
\begin{bmatrix} x_{\mathrm{ref},n+1} - k_x(x_{\mathrm{ref},n} - x_n) \\ y_{\mathrm{ref},n+1} - k_y(y_{\mathrm{ref},n} - y_n) \\ \psi_{\mathrm{ref},n+1} - k_\psi(\psi_{\mathrm{ref},n} - \psi_n) \\ u_{\mathrm{ref},n+1} - k_u(u_{\mathrm{ref},n} - u_n) \\ \omega_{\mathrm{ref},n+1} - k_\omega(\omega_{\mathrm{ref},n} - \omega_n) \end{bmatrix} = \begin{bmatrix} x_n \\ y_n \\ \psi_n \\ u_n \\ \omega_n \end{bmatrix} + T\left\{ \begin{bmatrix} u_{\mathrm{ref},n} \cos\psi_{\mathrm{ref},n} \\ u_{\mathrm{ref},n} \sin\psi_{\mathrm{ref},n} \\ \omega_{\mathrm{ref},n} \\ \frac{\theta_3}{\theta_1}\omega_n^2 - \frac{\theta_4}{\theta_1}u_n \\ -\frac{\theta_5}{\theta_2}u_n\omega_n - \frac{\theta_6}{\theta_2}\omega_n \end{bmatrix} \right.
$$

$$
\left. \times + \begin{bmatrix} 0 & 0 \\ 0 & 0 \\ 0 & 0 \\ \frac{1}{\theta_1} & 0 \\ 0 & \frac{1}{\theta_2} \end{bmatrix} \begin{bmatrix} u_{c,n} \\ \omega_{c,n} \end{bmatrix} \right\} \tag{4.33}
$$

Now, the conditions such that (4.33) has exact solution are obtained. Thus, considering the first three rows of system (4.33), it can be written:

$$
\begin{bmatrix} \cos{}_{\mathrm{ref},n} & 0 \\ \sin\psi_{\mathrm{ref},n} & 0 \\ 0 & 1 \end{bmatrix} \begin{bmatrix} u_{\mathrm{ref},n} \\ \omega_{\mathrm{ref},n} \end{bmatrix} = \frac{1}{T} \begin{bmatrix} x_{\mathrm{ref},n+1} - k_x(x_{\mathrm{ref},n} - x_n) - x_n \\ y_{\mathrm{ref},n+1} - k_y(y_{\mathrm{ref},n} - y_n) - y_n \\ \psi_{\mathrm{ref},n+1} - k_\psi(\psi_{\mathrm{ref},n} - \psi_n) - \psi_n \end{bmatrix} \tag{4.34}
$$

The sacrificed values of variables u and ω can be obtained, analyzing the conditions to have an exact solution for system (4.34). From the first two rows, the first condition for obtaining a unique solution is:

$$
\frac{\sin\psi_{\mathrm{ref},n}}{\cos\psi_{\mathrm{ref},n}} = \tan\psi_{\mathrm{ref},n} = \frac{y_{\mathrm{ref},n+1} - k_y(y_{\mathrm{ref},n} - y_n) - y_n}{x_{\mathrm{ref},n+1} - k_x(x_{\mathrm{ref},n} - x_n) - x_n} \tag{4.35}
$$

The values of these sacrificed variables are computed by solving (4.34) by least square

$$
\begin{bmatrix} u_{\mathrm{ref},n} \\ \omega_{\mathrm{ref},n} \end{bmatrix} = \frac{1}{T} \begin{bmatrix} (x_{\mathrm{ref},n+1} - k_x(x_{\mathrm{ref},n} - x_n) - x_n)\cos\psi_{\mathrm{ref},n} + (y_{\mathrm{ref},n+1} - k_y(y_{\mathrm{ref},n} - y_n) - y_n)\sin\psi_{\mathrm{ref},n} \\ \psi_{\mathrm{ref},n+1} - k_\psi(\psi_{\mathrm{ref},n} - \psi_n) - \psi_n \end{bmatrix} \tag{4.36}
$$

The value of $\psi_{\mathrm{ref},\, n\, +\, 1}$ is required to compute (4.36). This value can be expressed by the series:

$$\psi_{\text{ref},n+1} = \psi_{\text{ref}}|_{t=nT} + \frac{d}{dt}(\psi_{\text{ref}})|_{t=nT}T + \frac{d^2}{dt^2}(\psi_{\text{ref}})|_{t=nT}\frac{T^2}{2} + \cdots + \text{C.T.} \tag{4.37}$$

This value can be approximated by truncating the series. If only the first term is considered:

$$\psi_{\text{ref},n+1} \approx \psi_{\text{ref},n} \tag{4.38}$$

If two terms of the series are considered,

$$\begin{aligned} &\frac{d}{dt}(\psi_{\text{ref}})|_{t=nT} \approx \frac{\psi_{\text{ref},n} - \psi_{\text{ref},n-1}}{T} \Rightarrow \psi_{\text{ref},n+1} \approx \psi_{\text{ref}}|_{t=nT} + \frac{d}{dt}(\psi_{\text{ref}})|_{t=nT}T \\ &\psi_{\text{ref},n+1} \approx \psi_{\text{ref},n} + \frac{\psi_{\text{ref},n} - \psi_{\text{ref},n-1}}{T}T = 2\psi_{\text{ref},n} - \psi_{\text{ref},n-1} \end{aligned} \tag{4.39}$$

Finally, the control action values are obtained from system (4.32) by least square:

$$\begin{bmatrix} u_{c,n} \\ \omega_{c,n} \end{bmatrix} = \begin{bmatrix} \theta_1\left(\dfrac{(u_{\text{ref},n+1} - k_u(u_{\text{ref},n} - u_n) - u_n)}{T} - \left(\dfrac{\theta_3}{\theta_1}{\omega^2}_n - \dfrac{\theta_4}{\theta_1}u_n\right)\right) \\ \theta_2\left(\dfrac{\omega_{\text{ref},n+1} - k_\omega(\omega_{\text{ref},n} - \omega_n) - \omega_n}{T} - \left(-\dfrac{\theta_5}{\theta_2}u_n\omega_n - \dfrac{\theta_6}{\theta_2}\omega_n\right)\right) \end{bmatrix} \tag{4.40}$$

where $0 < k_x, k_y, k_\psi, k_u, k_\omega < 1$ and $e_{x,\,n} = (x_{\text{ref},\,n} - x_n)$; $e_{y,\,n} = (y_{\text{ref},\,n} - y_n)$; $e_{\psi,\,n} = (\psi_{\text{ref},\,n} - \psi_n)$; $e_{u,\,n} = (u_{\text{ref},\,n} - u_n)$; $e_{\omega,\,n} = (\omega_{\text{ref},\,n} - \omega_n)$.

4.2.2 Dynamic Controller Performance

Following a similar reasoning, the tracking errors are computed.

Theorem 4.3 If the system behavior is modeled by (4.32) and the controller is designed by (4.40), the tracking errors in following a given trajectory are such that $e_n \to 0$, $n \to \infty$ if the controller parameters are $0 < k_x < 1$, $0 < k_y < 1$, $0 < k_\psi < 1$, $0 < k_u < 1$, and $0 < k_\omega < 1$.

Proof In this case, by making the difference between system models (4.32) and (4.33), it yields

$$\begin{bmatrix} x_{\mathrm{ref},n+1} - x_{n+1} - k_x(x_{\mathrm{ref},n} - x_n) \\ y_{\mathrm{ref},n+1} - y_{n+1} - k_y(y_{\mathrm{ref},n} - y_n) \\ \psi_{\mathrm{ref},n+1} - \psi_{n+1} - k_\psi(\psi_{\mathrm{ref},n} - \psi_n) \\ u_{\mathrm{ref},n+1} - u_{n+1} - k_u(u_{\mathrm{ref},n} - u_n) \\ \omega_{\mathrm{ref},n+1} - \omega_{n+1} - k_\omega(\omega_{\mathrm{ref},n} - \omega_n) \end{bmatrix} = T \begin{bmatrix} u_{\mathrm{ref},n}\cos\psi_{\mathrm{ref},n} - u_n\cos\psi_n \\ u_{\mathrm{ref},n}\sin\psi_{\mathrm{ref},n} - u_n\sin\psi_n \\ \omega_{\mathrm{ref},n} - \omega_n \\ 0 \\ 0 \end{bmatrix} \quad (4.41)$$

After simple operations, the following can be obtained:

$$\begin{bmatrix} e_{x,n+1} - k_x e_{x,n} \\ e_{y,n+1} - k_y e_{y,n} \\ e_{\psi,n+1} - k_\psi e_{\psi,n} \\ e_{u,n+1} - k_u e_{u,n} \\ e_{\omega,n+1} - k_\omega e_{\omega,n} \end{bmatrix} = T \begin{bmatrix} u_{\mathrm{ref},n}\cos\psi_{\mathrm{ref},n} - u_n\cos\psi_n \\ u_{\mathrm{ref},n}\sin\psi_{\mathrm{ref},n} - u_n\sin\psi_n \\ e_{\omega,n} \\ 0 \\ 0 \end{bmatrix} \quad (4.42)$$

That is,

$$\begin{bmatrix} e_{x,n+1} \\ e_{y,n+1} \\ e_{\psi,n+1} \\ e_{u,n+1} \\ e_{\omega,n+1} \end{bmatrix} = \begin{bmatrix} k_x & 0 & 0 & 0 & 0 \\ 0 & k_y & 0 & 0 & 0 \\ 0 & 0 & k_\psi & 0 & 0 \\ 0 & 0 & 0 & k_u & 0 \\ 0 & 0 & 0 & 0 & k_\omega \end{bmatrix} \begin{bmatrix} e_{x,n} \\ e_{y,n} \\ e_{\psi,n} \\ e_{u,n} \\ e_{\omega,n} \end{bmatrix} + T \begin{bmatrix} u_{\mathrm{ref},n}\cos\psi_{\mathrm{ref},n} - u_n\cos\psi_n \\ u_{\mathrm{ref},n}\sin\psi_{\mathrm{ref},n} - u_n\sin\psi_n \\ e_{\omega,n} \\ 0 \\ 0 \end{bmatrix} \quad (4.43)$$

From (4.43), the convergence of the tracking errors can be established as follows:

$$\begin{bmatrix} e_{u,n} \\ e_{\omega,n} \end{bmatrix} = \begin{bmatrix} k_u e_{u,0} \\ k_\omega e_{\omega,0} \end{bmatrix} \Rightarrow \begin{bmatrix} e_{u,n} \\ e_{\omega,n} \end{bmatrix} \to \begin{bmatrix} 0 \\ 0 \end{bmatrix}, n \to \infty \quad (4.44)$$

Thence,

$$\begin{aligned} & e_{\psi,n+1} = k_\psi e_{\psi,n} + T e_{\omega,n} \\ & \text{if } |k_\psi| < 1, e_{\psi,n} \to 0, n \to \infty \\ & \Rightarrow \psi_n \to \psi_{\mathrm{ref},n}, n \to \infty \end{aligned} \quad (4.45)$$

Finally, as from (4.44) $u_n \to u_{\mathrm{ref},\,n}$, $n \to \infty$, the tracking error will be

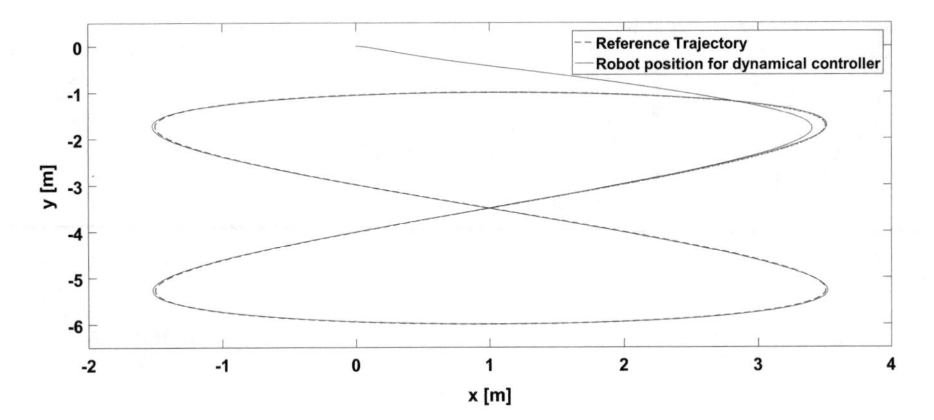

Fig. 4.10 Robot position and reference trajectory

$$\begin{bmatrix} u_n \sin\psi_n \\ u_n \cos\psi_n \end{bmatrix} \rightarrow \begin{bmatrix} u_{\text{ref},n} \sin\psi_{\text{ref},n} \\ u_{\text{ref},n} \cos\psi_{\text{ref},n} \end{bmatrix}, t \rightarrow \infty \Rightarrow \begin{bmatrix} e_{x,n} \\ e_{y,n} \end{bmatrix} \rightarrow \begin{bmatrix} 0 \\ 0 \end{bmatrix}, n \rightarrow \infty \qquad (4.46)$$

4.2.3 Experimental Results

Now the performance of the control law (4.40) is evaluated by using the same experiment used in the kinematic controllers developed following the Euler and trapezoidal models presented in the previous section. Here, the same robot, reference trajectory, and initial conditions of operation are used. Later, a comparative analysis is presented with all the results obtained when applied to the PIONEER mobile robot.

The controller parameters were selected to get a smooth trajectory tracking, ensuring that the tracking error tends to zero. Thus, with this consideration in mind, the controller parameters used in the experimentation with the dynamical controller are: $k_x = 0.95$, $k_y = 0.95$, $k_\psi = 0.9$, $k_u = 0.94$ and $k_\omega = 0.94$.

The robot position and the reference trajectory are shown in Fig. 4.10. As it can be seen, the robot reaches and follows the reference trajectory with good precision. Furthermore, the tracking errors shown in Fig. 4.11 are low and remain close to zero. The computed control actions and the robot velocities are shown in Fig. 4.12.

4.2.4 Comparative Analysis

In the earlier sections, three LAB controllers were developed and applied to a mobile robot, using the same initial conditions and reference trajectory. Now the results

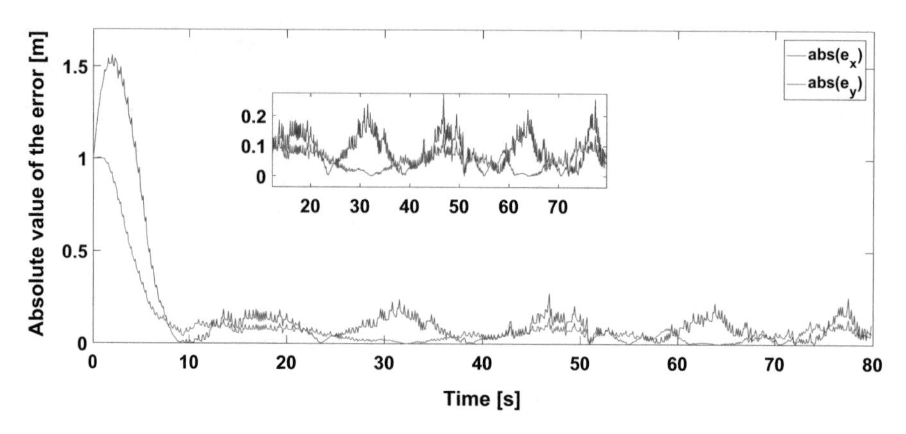

Fig. 4.11 Absolute value of the tracking errors

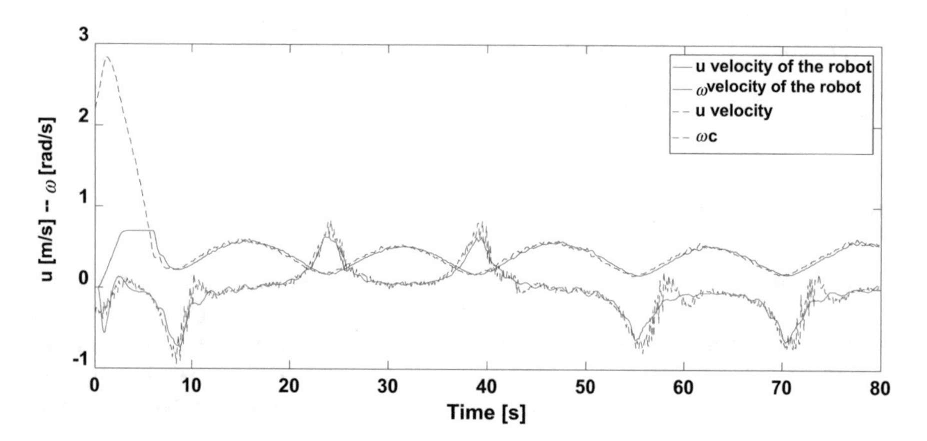

Fig. 4.12 Computed control actions and robot velocities

obtained with each controller are summarized in the same plots, in order to compare the performance obtained with each controller.

Figure 4.13 shows the position of the robot and the reference trajectory in all the performed experimentations. As it can be seen, all the controllers reach and follow the reference trajectory without undesired trajectory deviations. In addition, the trajectory achieved when any controller is used is smooth by a proper controller parameter selection. The tracking errors in y-coordinate and x-coordinate are shown in Figs. 4.14 and 4.15. As it can be seen, the dynamical controller presents better performance. This result was expected as the dynamical controller has been developed using a mathematical model, which is more approximated to the real system. This result can be confirmed when analyzing the Figs. 4.16 and 4.17 where the position of the robot in the y-coordinate and x-coordinate versus time are shown.

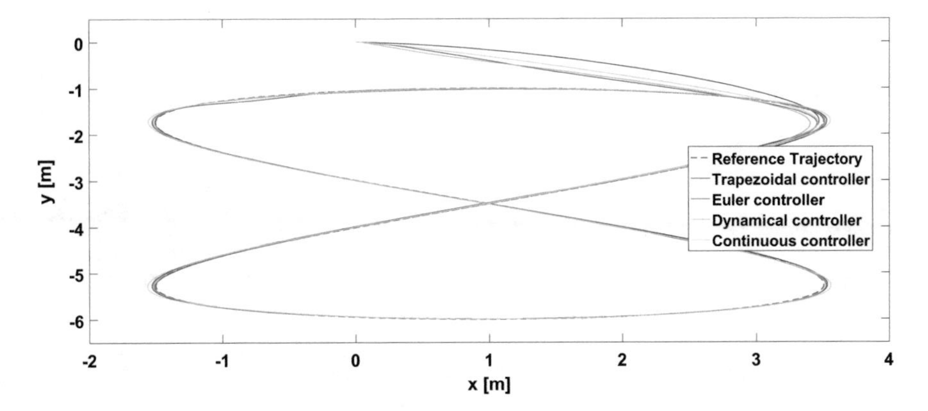

Fig. 4.13 Robot position and reference trajectory

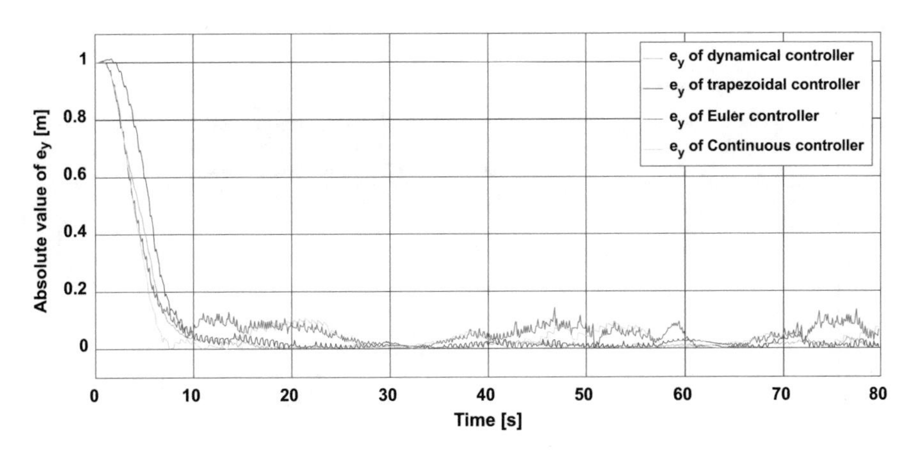

Fig. 4.14 Tracking error in the *y*-coordinate

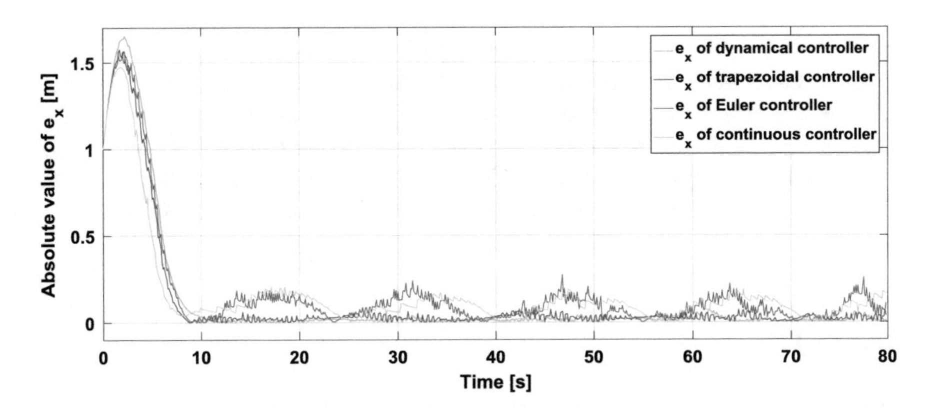

Fig. 4.15 Tracking error in the *x*-coordinate

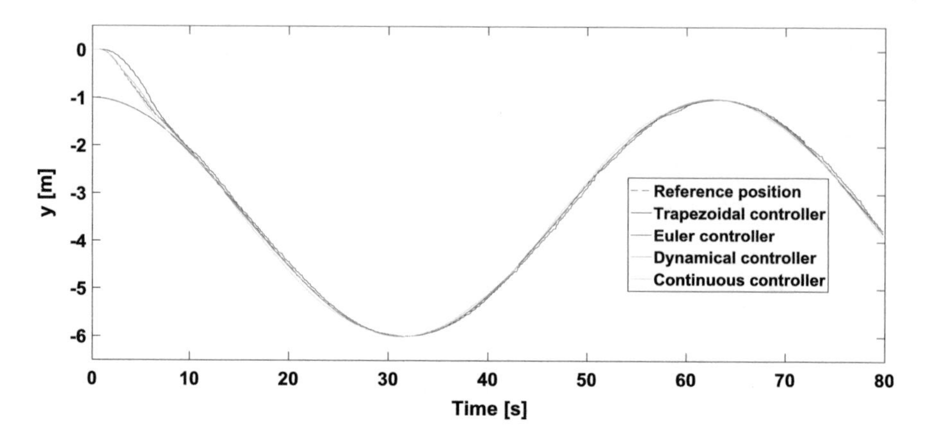

Fig. 4.16 Position of the robot in the y-coordinate

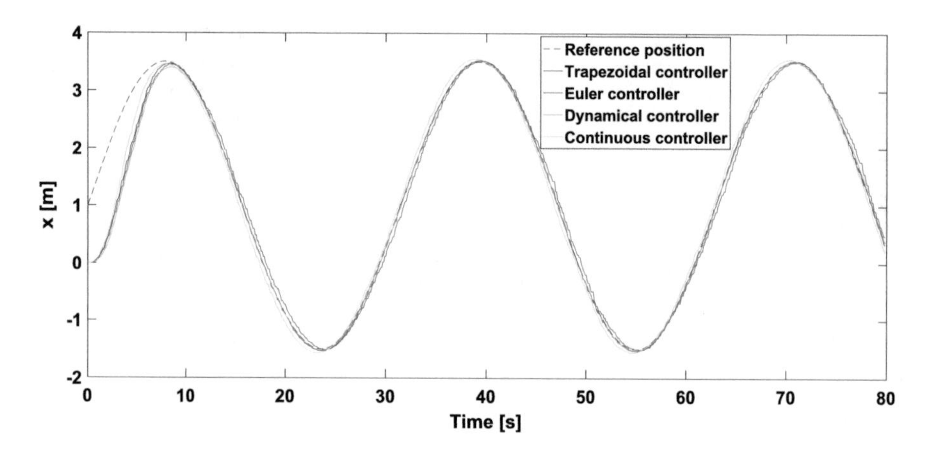

Fig. 4.17 Position of the robot in the x-coordinate

Remark 4.4 Usually, when a dynamic model of the robot is used to derive the control, the sequence is to design a control based on the kinematic model and then make a refinement to incorporate the dynamics (Capraro et al., 2017). If the LAB CD methodology is applied to the dynamic model (4.32), this sequence is implicit: the requirement of unique solution for (4.33) leads to the references for the sacrificed variables (4.35) and (4.36), and then the control actions are computed (4.40).

References

Capraro, F., Rossomando, F. G., Soria, C., & Scaglia, G. (2017). Cascade sliding control for trajectory tracking of a nonholonomic mobile robot with adaptive neural compensator. *Mathematical Problems in Engineering*.

De La Cruz, C., & Carelli, R. (2008). Dynamic model based formation control and obstacle avoidance of multi-robot systems. *Robotica, 26*(3), 345–356.

Rosales, A., Scaglia, G., Mut, V., & di Sciascio, F. (2009). Trajectory tracking of mobile robots in dynamic environments—a linear algebra approach. *Robotica, 27*(7), 981–997. https://doi.org/10.1017/S0263574709005402.

Scaglia, G., Quintero, O. L., Mut, V., & di Sciascio, F. (2008). Numerical methods based controller design for mobile robots. *IFAC Proceedings Volumes, 41*(2), 4820–4827.

Scaglia, G., Rosales, A., Quintero, L., Mut, V., & Agarwal, R. (2010). A linear-interpolation-based controller design for trajectory tracking of mobile robots. *Control Engineering Practice, 18*(3), 318–329.

Scaglia, G., Serrano, E., Rosales, A., & Albertos, P. (2019). Tracking control design in nonlinear multivariable systems: robotic applications. *Mathematical Problems in Engineering, 2019*, 8643515. https://doi.org/10.1155/2019/8643515.

Serrano, M. E., Godoy, S., Mut, V., Ortiz, O., & Scaglia, G. (2016). A nonlinear trajectory tracking controller for mobile robots with velocity limitation via parameters regulation. *Robotica, 34*(11), 2546–2565. https://doi.org/10.1017/S026357471500020X.

Serrano, M. E., Scaglia, G. J. E., Cheein, F. A., Mut, V., & Ortiz, O. A. (2015). Trajectory-tracking controller design with constraints in the control signals: a case study in mobile robots. *Robotica, 33*, 2186–2203. https://doi.org/10.1017/S0263574714001325.

Chapter 5
Application to Marine and Aerial Vehicles

As previously discussed, autonomous vehicles can be classified into three types: terrestrial, marine, and aerial. In the previous chapters, the control design methodology based on linear algebra was applied to mobile robots, one of the most studied terrestrial autonomous vehicles in the bibliography. In this chapter, the application of the methodology to marine and aerial vehicles will be extended. Here, three representative systems are considered to show that the Linear Algebra-Based Control Design methodology can be applied to systems of different nature.

First, a model of a marine vessel will be considered. In this application, the number of tracked variables is the same as that of the control actions, but the number of sacrificed variables is higher than required. Moreover, two of these sacrificed variables are determined by the same control input. Thus, one of them is useless to compute the control action, and its reference will not be considered. In any case, its dynamic evolution should be analyzed to ensure the stability of the whole plant.

As a second class of unmanned vehicles, two aerial autonomous devices are considered. First, the control of a planar vertical take-off and landing (PVTOL) aircraft is developed. The LAB CD methodology is applied and excellent results are obtained. Finally, the control of a quadrotor is considered. In this case, the state vector dimension is much higher, but the procedure is the same and a control solution is easily found. In all cases, the constraints on the model and the accessibility of the state are assumed to be fulfilled.

A Simulink diagram for the simulation of the control for the marine vessel is included at the end of the chapter. Similar diagrams can be developed by the reader, allowing a better understanding of the control structure as well as an easy tuning of the control parameters, to realize their influence in the controlled plant performance.

G. Scaglia et al., *Linear Algebra Based Controllers*,
https://doi.org/10.1007/978-3-030-42818-1_5

5.1 Application to Marine Vessels

In the past decades, there has been significant interest in the trajectory tracking control of autonomous marine vessels. These vehicles are widely used to inspection in dangerous areas, transport, and military applications. In this section, the design method based on linear algebra described in Chap. 2 is applied to control the trajectory of a marine vessel.

5.1.1 *Marine Vessel Model*

Consider the mathematical representation of the marine vessel shown in Fig. 5.1 described by the 3-DOF model (5.1), (see Børhaug, Pavlov, Panteley, & Pettersen, 2011; Serrano, Scaglia, Godoy, Mut, & Ortiz, 2013),

$$
\begin{aligned}
\dot{x} &= u\,\cos\psi - v\,\sin\psi \\
\dot{y} &= u\,\sin\psi + v\,\cos\psi \\
\dot{\psi} &= r \\
\dot{u} &= \frac{m_{22}}{m_{11}} vr + \frac{m_{23}}{m_{11}} r^2 - \frac{d_{11}}{m_{11}} u + \frac{b_{11}}{m_{11}} \tau_u \\
\dot{v} &= -\frac{m_{23}}{m_{22}} \dot{r} + \frac{m_{11}}{m_{22}} ur + \frac{d_{22}}{m_{22}} v + \frac{d_{23}}{m_{22}} r \\
\dot{r} &= -\frac{m_{23}}{m_{33}} \dot{v} + \frac{m_{11} - m_{22}}{m_{33}} vu - \frac{m_{23}}{m_{33}} ur - \frac{d_{32}}{m_{33}} v - \frac{d_{33}}{m_{33}} r + \frac{b_{32}}{m_{33}} \tau_r
\end{aligned}
\tag{5.1}
$$

where the vector $[x, y, \psi]^T$ represents the earth-fixed position and heading, and the vector $[u, v, r]^T$ represents the corresponding velocities. $[\tau_u, \tau_r]^T$ is the control input

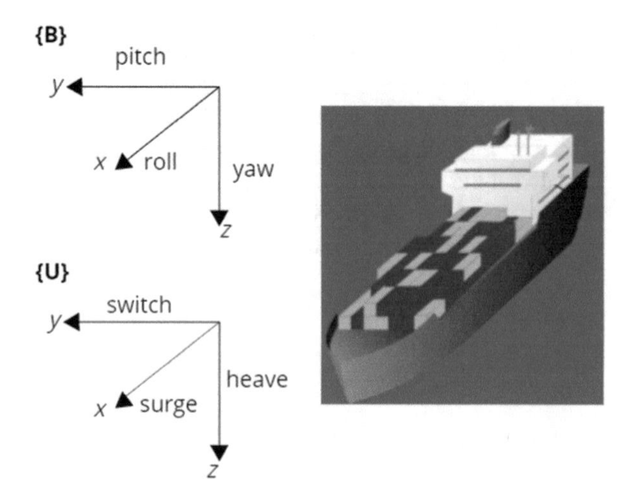

Fig. 5.1 Marine vessel: global coordinate frame {U} and body-fixed coordinate frame {B}

vector, where τ_u is the surge control, and τ_r is the yaw control, respectively. The meaning of the different parameters can be seen in the reference above.

In order to write the model in the internal representation form, the last two equations are combined to get

$$\begin{aligned}
\dot{x} &= u\ \cos\psi\text{-}v\ \sin\psi \\
\dot{y} &= u\ \sin\psi + v\ \cos\psi \\
\dot{\psi} &= r \\
\dot{u} &= \frac{m_{22}}{m_{11}}vr + \frac{m_{23}}{m_{11}}r^2 - \frac{d_{11}}{m_{11}}u + \frac{b_{11}}{m_{11}}\tau_u \\
\dot{v} &= a_{11}vu + a_{12}ur + a_{13}v + a_{14}r + \beta_1\tau_r \\
\dot{r} &= a_{21}vu + a_{22}ur + a_{23}v + a_{24}r + \beta_2\tau_r
\end{aligned} \tag{5.2}$$

where the new coefficients are related to the original parameters as follows

$$\begin{aligned}
a_{11} &= \frac{-m_{23}(m_{11}-m_{22})}{m_{22}m_{33}-m_{23}^2} & a_{21} &= \frac{m_{22}(m_{11}-m_{22})}{m_{22}m_{33}-m_{23}^2} \\
a_{12} &= \frac{m_{11}m_{33}+m_{23}^2}{m_{22}m_{33}-m_{23}^2} & a_{22} &= \frac{m_{11}m_{23}-m_{23}m_{22}}{m_{22}m_{33}-m_{23}^2} \\
a_{13} &= \frac{d_{22}m_{33}+d_{32}m_{23}}{m_{22}m_{33}-m_{23}^2} & a_{23} &= \frac{d_{22}m_{23}-d_{32}m_{22}}{m_{22}m_{33}-m_{23}^2} \\
a_{14} &= \frac{d_{23}m_{33}+d_{33}m_{22}}{m_{22}m_{33}-m_{23}^2} & a_{24} &= \frac{d_{23}m_{23}-d_{33}m_{22}}{m_{22}m_{33}-m_{23}^2} \\
\beta_1 &= \frac{-m_{32}b_{32}}{m_{22}m_{33}-m_{23}^2} & \beta_2 &= \frac{m_{22}b_{32}}{m_{22}m_{33}-m_{23}^2}
\end{aligned} \tag{5.3}$$

As the control is going to be implemented in DT, a sampling period T is assumed, and the discrete representation of system (5.2) is obtained by applying an Euler discretization approximation:

$$\begin{bmatrix} x_{n+1} \\ y_{n+1} \\ \psi_{n+1} \\ u_{n+1} \\ v_{n+1} \\ r_{n+1} \end{bmatrix} = \begin{bmatrix} x_n \\ y_n \\ \psi_n \\ u_n \\ v_n \\ r_n \end{bmatrix} + T\begin{bmatrix} u_n\ \cos\psi_n - v_n\ \sin\psi_n \\ u_n\ \sin\psi_n + v_n\ \cos\psi_n \\ r_n \\ -\frac{m_{22}}{m_{11}}v_n r_n - \frac{m_{23}}{m_{11}}r_n^2 + \frac{d_{11}}{m_{11}}u_n \\ f_{1,n} \\ f_{2,n} \end{bmatrix} + T\begin{bmatrix} 0 \\ 0 \\ 0 \\ \frac{b_{11}}{m_{11}}\tau_{u,n} \\ \beta_1\tau_{r,n} \\ \beta_2\tau_{r,n} \end{bmatrix} \tag{5.4}$$

where $f_{i,n}$ $(i = 1,2)$ is defined to simplify the system representation being:

$$\begin{aligned} f_{1n} &= a_{11}v_n u_n + a_{12}r_n u_n + a_{13}v_n + a_{14}r_n \\ f_{2n} &= a_{21}v_n u_n + a_{22}r_n u_n + a_{23}v_n + a_{24}r_n \end{aligned} \tag{5.5}$$

5.1.2 Controller Design

The control objective is to find the combined control actions τ_u and τ_r that must be applied such that the ship reaches and follows a desired trajectory $\xi_{\text{ref}} = [x_{\text{ref}}\ y_{\text{ref}}]^T$. Observe that the remaining state variables will be assumed as sacrificed variables, $z = [\psi u v r]^T$, their reference being flexible to allow a smooth trajectory tracking.

Following the LAB CD technique, the controlled system will approach the references as defined in (2.29), that is,

$$\begin{bmatrix} \xi_{n+1} \\ z_{n+1} \end{bmatrix} = \begin{bmatrix} \xi_{\text{ref},n+1} \\ z_{\text{ref},n+1} \end{bmatrix} - \begin{bmatrix} k_\xi(\xi_{\text{ref},n} - \xi_n) \\ k_z(z_{\text{ref},n} - z_n) \end{bmatrix} \tag{5.6}$$

where not all the sacrificed variables will require a reference, and the controlled plant model will be such as (2.28), that is,

$$\begin{bmatrix} x_{\text{ref},n+1} - k_x(x_{\text{ref},n} - x_n) - x_n - T(u_n \cos\psi_n - v_n \sin\psi_n) \\ y_{\text{ref},n+1} - k_y(y_{\text{ref},n} - y_n) - y_n - T(u_n \sin\psi_n + v_n \cos\psi_n) \\ \psi_{\text{ref},n+1} - k_\psi(\psi_{\text{ref},n} - \psi_n) - \psi_n - Tr_n \\ u_{\text{ref},n+1} - k_u(u_{\text{ref},n} - u_n) - u_n - T\left(-\frac{m_{22}}{m_{11}}v_n r_n - \frac{m_{23}}{m_{11}}r_n^2 + \frac{d_{11}}{m_{11}}u_n\right) \\ v_{\text{ref},n+1} - k_v(v_{\text{ref},n} - v_n) - v_n - Tf_{1n} \\ r_{\text{ref},n+1} - k_r(r_{\text{ref},n} - r_n) - r_n - Tf_{2n} \end{bmatrix}$$
$$= \begin{bmatrix} 0 & 0 \\ 0 & 0 \\ 0 & 0 \\ T\frac{b_{11}}{m_{11}} & 0 \\ 0 & T\beta_1 \\ 0 & T\beta_2 \end{bmatrix} \begin{bmatrix} \tau_{u,n} \\ \tau_{r,n} \end{bmatrix} \tag{5.7}$$

which can be written as

$$\underbrace{\begin{bmatrix} \frac{\Delta_{x,n}}{T} - u_n \cos\psi_n + v_n \sin\psi_n \\ \frac{\Delta_{y,n}}{T} - u_n \sin\psi_n - v_n \cos\psi_n \\ \frac{\Delta_{\psi,n}}{T} - r_n \\ m_{11}\frac{\Delta_{u,n}}{T} - m_{22}v_n r_n - m_{23}r_n^2 + d_{11}u_n \\ \frac{\Delta_{v,n}}{T} - \frac{f_{1,n}}{T} \\ \frac{\Delta_{r,n}}{T} - \frac{f_{2,n}}{T} \end{bmatrix}}_{b} = \underbrace{\begin{bmatrix} 0 & 0 \\ 0 & 0 \\ 0 & 0 \\ b_{11} & 0 \\ 0 & \beta_1 \\ 0 & \beta_2 \end{bmatrix}}_{A} \underbrace{\begin{bmatrix} \tau_{u,n} \\ \tau_{r,n} \end{bmatrix}}_{\tau} \tag{5.8}$$

where

$$\begin{aligned} \Delta_{x,n} &= x_{\text{ref},n+1} - k_x\left(x_{\text{ref},n} - x_n\right) - x_n & \Delta_{y,n} &= y_{\text{ref},n+1} - k_y\left(y_{\text{ref},n} - y_n\right) - y_n \\ \Delta_{\psi,n} &= \psi_{\text{ref},n+1} - k_\psi\left(\psi_{\text{ref},n} - \psi_n\right) - \psi_n & \Delta_{u,n} &= u_{\text{ref},n+1} - k_u\left(u_{\text{ref},n} - u_n\right) - u_n \\ \Delta_{v,n} &= v_{\text{ref},n+1} - k_v\left(v_{\text{ref},n} - v_n\right) - v_n & \Delta_{r,n} &= r_{\text{ref},n+1} - k_r\left(r_{\text{ref},n} - r_n\right) - r_n \end{aligned} \tag{5.9}$$

The next step is to find the conditions that must be accomplished by the reference of the sacrificed variables, in order that system (5.8) has an exact solution. Looking at the elements of A, it is clear that the first three elements of b should vanish, whereas the last two sacrificed variables are determined by the same control input, τ_r. Thus, the reference for only one sacrificed variable is needed and the reference for v will be omitted.

Going step by step, consider the first two rows of (5.8) to determine the reference for the orientation, as well as the reference for the velocity u. In this setting, the reference for the other sacrificed variables is irrelevant, and they will not be considered. Thus:

$$\underbrace{\begin{bmatrix} \cos\psi_{\text{ref},n} \\ \sin\psi_{\text{ref},n} \end{bmatrix}}_{P} \underbrace{u_{\text{ref},n}}_{s} = \underbrace{\begin{bmatrix} \frac{\Delta_{x,n}}{T} + v_n \sin\psi_n \\ \frac{\Delta_{y,n}}{T} - v_n \cos\psi_n \end{bmatrix}}_{q} \tag{5.10}$$

In order to have an exact solution in system (5.10), the orientation angle reference must fulfill that:

$$\tan \psi_{\text{ref},n} = \frac{\sin \psi_{\text{ref},n}}{\cos \psi_{\text{ref},n}} = \frac{\frac{\Delta_{y,n}}{T} - v_n \cos \psi_n}{\frac{\Delta_{x,n}}{T} + v_n \sin \psi_n} \tag{5.11}$$

This value represents the necessary orientation that must be adopted by the sacrificed variable ψ_{ref}. This orientation value ensures that the ship reaches and follows the reference trajectory. Then, the value that must be taken by the sacrificed variable $u_{\text{ref},n}$ is calculated solving system (5.10), replacing the ship orientation by the reference given by (5.11):

$$u_{\text{ref},n} = \left(\frac{\Delta_{x,n}}{T} + v_n \sin \psi_n\right) \cos \psi_{\text{ref},n} + \left(\frac{\Delta_{y,n}}{T} - v_n \cos \psi_n\right) \sin \psi_{\text{ref},n} \tag{5.12}$$

The third condition to have a unique solution in system (5.8) is analyzed, considering the third row. In this step, the value of the reference $r_{\text{ref},n}$ naturally arises:

$$r_{\text{ref},n} = \frac{\Delta_{\psi,n}}{T} \tag{5.13}$$

where $\Delta_{\psi,\, n}$ has been defined in (5.9). Based on the sequence of previous values for the orientation reference, $\{\psi_{\text{ref},i}\}$for $i = 0, 1, \cdots, n$, the next value $\psi_{\text{ref},\, n+1}$ required in (5.13) should be estimated. A similar treatment will be used for computing other increments of sacrificed variables, like $\Delta_{r,\, n}$ or $\Delta_{u,\, n}$, in the following computations. In general, it is reasonable to assume a linear progression of the references for the sacrificed variables, as they are just tentative to determine the solvability of (5.8).

It should be noticed that, as already discussed, in order to compute the references for the two first sacrificed variables, in (5.11) and (5.12), the next values of the trajectory reference should be known but, for the last remaining sacrificed variables, (5.13), the value of the reference for the sacrificed variables in the next sampling time should be estimated.

The references for the sacrificed variables are then replaced in (5.8), and the control actions that must be applied at each sample time are obtained by solving (5.8) by least square. In order to handle a full rank A-submatrix, two rows dealing independently with the control actions should be selected. If rows fourth and sixth are chosen, the subsystem will be

$$\begin{bmatrix} b_{11} & 0 \\ 0 & \beta_2 \end{bmatrix} \begin{bmatrix} \tau_{u,n} \\ \tau_{r,n} \end{bmatrix} = \begin{bmatrix} m_{11}\frac{\Delta_{u,n}}{T} - m_{22} v_n r_n - m_{23} r_n^2 + d_{11} u_n \\ \frac{\Delta_{r,n} - f_{2,n}}{T} \end{bmatrix} \tag{5.14}$$

and the control actions are obtained

$$\begin{bmatrix} \tau_{u,n} \\ \tau_{r,n} \end{bmatrix} = \begin{bmatrix} \frac{1}{b_{11}}\left(m_{11}\frac{\Delta_{u,n}}{T} - m_{22}v_n r_n - m_{23}r_n^2 + d_{11}u_n\right) \\ \frac{1}{\beta_2}\frac{\Delta_{r,n} - f_{2,n}}{T} \end{bmatrix} \tag{5.15}$$

It should be noticed that, in this case, the reference for the v_n variable does not need to be computed, the actual value of this variable being determined by the computed control actionτ_r.

5.1.3 *Controlled Plant Behavior*

In order to analyze the performance of the controlled plant, the behavior of the system described by the model (5.4) and the control signal computed as in (5.15) should be determined. First, the behavior of the sacrificed variables is considered.

Combining the appropriate rows in (5.4) with the control action given by (5.15), and taking into account (5.9), the controlled plant model becomes

$$\begin{bmatrix} u_{n+1} \\ r_{n+1} \end{bmatrix} = \begin{bmatrix} u_{\text{ref},n+1} - k_u\left(u_{\text{ref},n} - u_n\right) \\ r_{\text{ref},n+1} - k_r\left(r_{\text{ref},n} - r_n\right) \end{bmatrix} \tag{5.16}$$

Thus, the errors in these two sacrificed variables follow the dynamics given by

$$\begin{bmatrix} e_{u,n+1} \\ e_{r,n+1} \end{bmatrix} = \begin{bmatrix} k_u & 0 \\ 0 & k_r \end{bmatrix} \begin{bmatrix} e_{u,n} \\ e_{r,n} \end{bmatrix} \tag{5.17}$$

If the absolute value of the controller parameters is less than one, these errors driven by the control actions (5.15) tend to zero. Now, the errors in the other sacrificed variables are considered. Looking at the third row in (5.4) and the previous error

$$\left.\begin{aligned} \psi_{n+1} &= \psi + Tr_n \\ r_n &= r_{\text{ref},n} - e_{r,n} \end{aligned}\right\} \Rightarrow \psi_{n+1} = \psi + T(r_{\text{ref},n} - e_{r,n}) \tag{5.18}$$

The reference $r_{\text{ref},\,n}$ has been computed by (5.13). Thence,

$$\psi_{n+1} = \psi_{\text{ref},n+1} - k_\psi\left(\psi_{\text{ref},n} - \psi_n\right) - Te_{r,n} \tag{5.19}$$

$$\Downarrow$$

$$e_{\psi,n+1} = k_\psi e_{\psi,n} + Te_{r,n} \tag{5.20}$$

Combining (5.17) and (5.20), the errors in the sacrificed variables can be written as

$$\begin{bmatrix} e_{\psi,n+1} \\ e_{u,n+1} \\ e_{r,n+1} \end{bmatrix} = \begin{bmatrix} k_\psi & 0 & T \\ 0 & k_u & 0 \\ 0 & 0 & k_r \end{bmatrix} \begin{bmatrix} e_{\psi,n} \\ e_{u,n} \\ e_{r,n} \end{bmatrix} \tag{5.21}$$

Thence, the last two sacrificed variables' error tends to zero driven by the control inputs, dragging also the third sacrificed variable.

In a similar way, the tracking errors in following the plant reference go to zero. From (5.4), the evolution of the main variables is determined by

$$\begin{bmatrix} x_{n+1} \\ y_{n+1} \end{bmatrix} = \begin{bmatrix} x_n \\ y_n \end{bmatrix} + T \begin{bmatrix} \cos\psi_n \\ \sin\psi_n \end{bmatrix} u_n + T \begin{bmatrix} -\sin\psi_n \\ \cos\psi_n \end{bmatrix} v_n \tag{5.22}$$

and the reference of the sacrificed variables $u_{\text{ref},\,n}$ and $\psi_{\text{ref},\,n}$ are computed in (5.10) to allow for an exact solution of

$$\begin{bmatrix} x_{\text{ref},n+1} - k_x\left(x_{\text{ref},n} - x_n\right) \\ y_{\text{ref},n+1} - k_y\left(y_{\text{ref},n} - y_n\right) \end{bmatrix} = \begin{bmatrix} x_n \\ y_n \end{bmatrix} + T \begin{bmatrix} \cos\psi_{\text{ref},n} \\ \sin\psi_{\text{ref},n} \end{bmatrix} u_{\text{ref},n} + T \begin{bmatrix} -\sin\psi_n \\ \cos\psi_n \end{bmatrix} v_n \tag{5.23}$$

From (5.22), (5.23)

$$\begin{bmatrix} x_{\text{ref},n+1} - x_{n+1} - k_x\left(x_{\text{ref},n} - x_n\right) \\ y_{\text{ref},n+1} - y_{n+1} - k_y\left(y_{\text{ref},n} - y_n\right) \end{bmatrix} = T \begin{bmatrix} \cos\psi_{\text{ref},n} \\ \sin\psi_{\text{ref},n} \end{bmatrix} u_{\text{ref},n} - T \begin{bmatrix} \cos\psi_n \\ \sin\psi_n \end{bmatrix} u_n \tag{5.24}$$

that is,

$$\begin{bmatrix} e_{x,n+1} \\ e_{y,n+1} \end{bmatrix} = \begin{bmatrix} k_x & 0 \\ 0 & k_y \end{bmatrix} \begin{bmatrix} e_{x,n} \\ e_{y,n} \end{bmatrix} + \underbrace{T \begin{bmatrix} \cos\psi_{\text{ref},n} \\ \sin\psi_{\text{ref},n} \end{bmatrix} u_{\text{ref},n} - T \begin{bmatrix} \cos\psi_n \\ \sin\psi_n \end{bmatrix} u_n}_{h\left(e_{\psi,n},\, e_{u,n}\right)} \tag{5.25}$$

Thus, the error dynamics is determined by a linear term plus an extra nonlinear term, $h(e_{\psi,\,n}, e_{u,\,n})$, which also tends to zero as far as the arguments $\{e_{\psi,\,n}, e_{u,\,n}\}$ tend to zero. In fact, applying the Taylor interpolation rule to $\cos\psi_n$, the following expression is obtained:

$$\cos\psi_n = \cos\psi_{\text{ref},n} + \sin\left(\underbrace{\psi_{\text{ref},n} + \lambda_1(\psi_n - \psi_{\text{ref},n})}_{\psi_1}\right)e_{\psi,n}; \quad 0 < \lambda_1 < 1$$

where ψ_1 is an intermediate angle between ψ_n y $\psi_{\text{ref},\,n}$. Also, for sin ψ_n

$$\sin\psi_n = \sin\psi_{\text{ref},n} - \cos\left(\underbrace{\psi_{\text{ref},n} + \lambda_2(\psi_n - \psi_{\text{ref},n})}_{\psi_2}\right)e_{\psi,n}; \quad 0 < \lambda_2 < 1$$

It is worth to note that both sin ψ_1 and cos ψ_2 are bounded. Substituting these terms in (5.25), it yields

$$\begin{aligned} h(e_{\psi,n}, e_{u,n}) &= T\begin{bmatrix} \cos\psi_{\text{ref},n} \\ \sin\psi_{\text{ref},n} \end{bmatrix} u_{\text{ref},n} \\ &- T\begin{bmatrix} \cos\psi_{\text{ref},n} + \sin\left(\underbrace{\psi_{\text{ref},n} + \lambda_1(\psi_n - \psi_{\text{ref},n})}_{\psi_1}\right)e_{\psi,n} \\ \sin\psi_{\text{ref},n} - \cos\left(\underbrace{\psi_{\text{ref},n} + \lambda_2(\psi_n - \psi_{\text{ref},n})}_{\psi_2}\right)e_{\psi,n} \end{bmatrix} \\ &\times (u_{\text{ref},n} - e_{u,n}) \end{aligned} \tag{5.26}$$

Thus,

$$\begin{bmatrix} e_{x,n+1} \\ e_{y,n+1} \end{bmatrix} = \begin{bmatrix} k_x & 0 \\ 0 & k_y \end{bmatrix} \times \begin{bmatrix} e_{x,n} \\ e_{y,n} \end{bmatrix} \underbrace{- T\begin{bmatrix} \sin\psi_1 u_{\text{ref},n} \\ -\cos\psi_2 u_{\text{ref},n} \end{bmatrix} e_{\psi,n} + T\begin{bmatrix} \cos\psi_{\text{ref},n} \\ \sin\psi_{\text{ref},n} \end{bmatrix} e_{u,n} + T\begin{bmatrix} \sin\psi_1 \\ -\cos\psi_2 \end{bmatrix} e_{\psi,n}e_{u,n}}_{h(e_{\psi,n}, e_{u,n})} \tag{5.27}$$

Thence,

$$\lim_{n\to\infty} e_{\psi,n}, e_{u,n} = 0; \Rightarrow \left\| h(e_{\psi,n}, e_{u,n}) \right\| \to 0, n \to \infty; \ \Rightarrow \begin{bmatrix} e_{x,n} \\ e_{y,n} \end{bmatrix} \to \begin{bmatrix} 0 \\ 0 \end{bmatrix}, n \to \infty \tag{5.28}$$

5.1.4 Simulation Results

To validate the theoretical results, a simulation is carried out, using the programming environment Simulink platform of MatLab®. The control structure is as depicted in the Fig. 5.2.

The ship model parameters are obtained from Serrano et al., 2013, and the used sampling period was $T = 0.1$ s. A sinusoidal trajectory is generated with a forward velocity of $u = 0.2$ m/s. The reference trajectory starts at $(x_{ref}(0), y_{ref}(0)) = (1$ m, 2 m$)$ and the initial position of the ship is at the system origin. The controller parameters are chosen as:

$$k_x = 0.9 \quad k_y = 0.94 \quad k_\psi = 0.31 \quad k_u = 0.78 \quad k_r = 0.88.$$

The system performance is shown in Fig. 5.3. As it can be seen in Fig. 5.4, the ship reaches the reference trajectory quickly and then continues without undesirable oscillations. The time variation of the absolute values of the tracking errors is shown in Fig. 5.5. They are bounded and remain close to zero. The control actions versus time are shown in Figs. 5.6 and 5.7, respectively.

Remark 5.1 At the end of this chapter, in Appendix 5.1, the Simulink diagram of the vessel model described here is detailed. The reader may implement this diagram

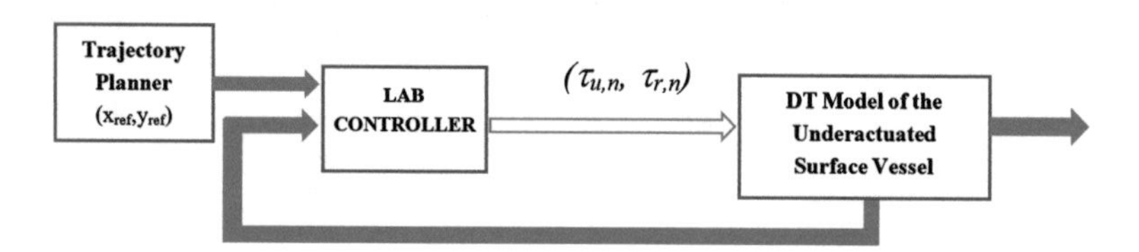

Fig. 5.2 Control structure

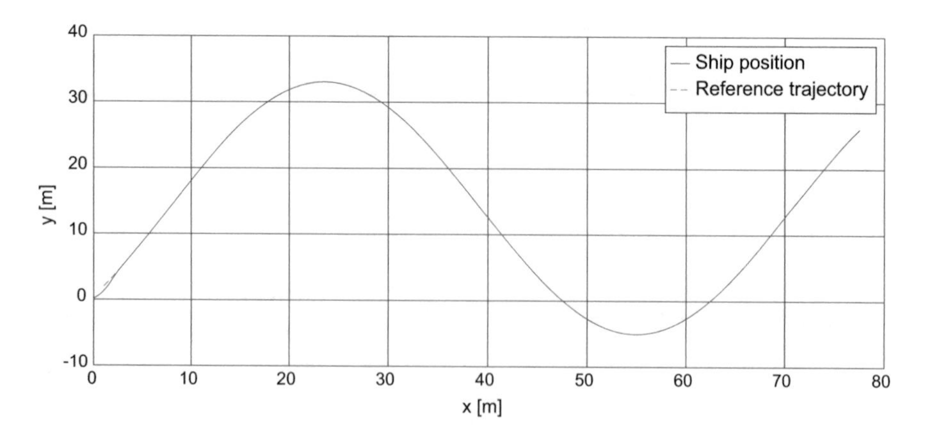

Fig. 5.3 Trajectory tracking of marine vessel in a sinusoidal reference

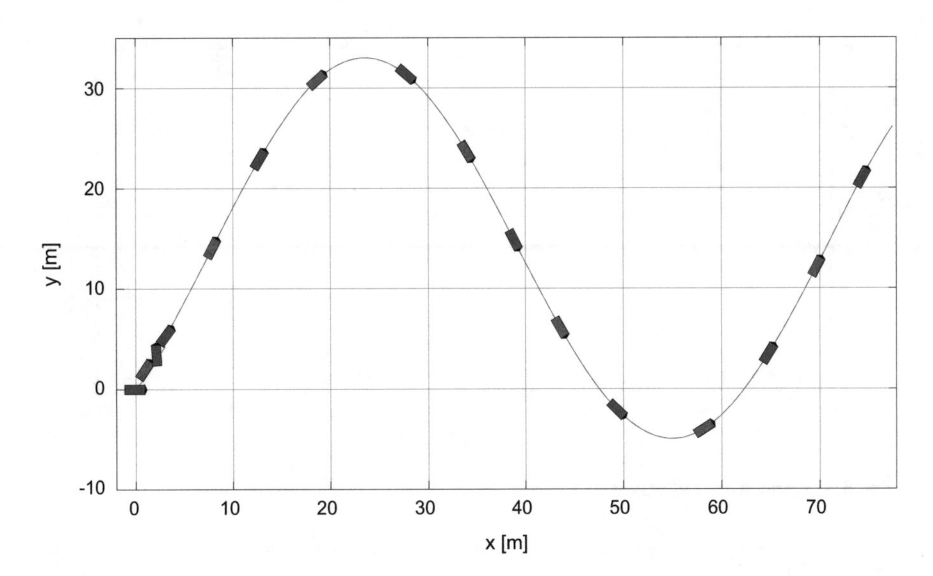

Fig. 5.4 Ship position following the sinusoidal reference, with a plot of the ship orientation

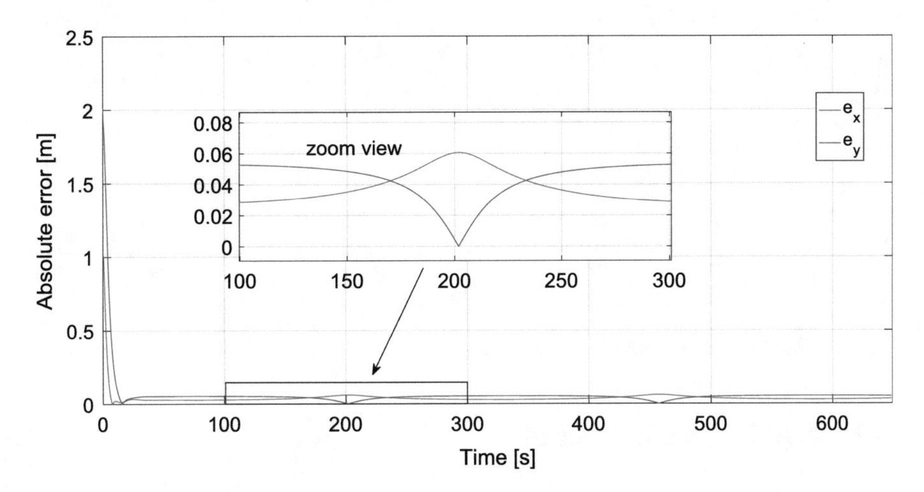

Fig. 5.5 Absolute values of the tracking errors versus time

and reproduce the earlier figures. Also, by changing the controller parameters the reader may experience about the simplicity of their selection. Other than that, the robustness to external disturbances or changes in the model parameters can also be verified by the reader, to get a better insight into the LAB CD methodology.

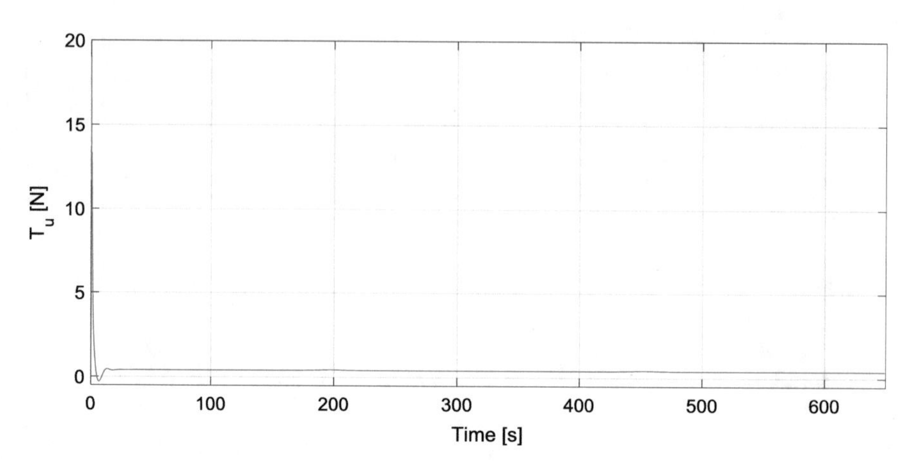

Fig. 5.6 Applied control action $\boldsymbol{\tau}_u$ versus time

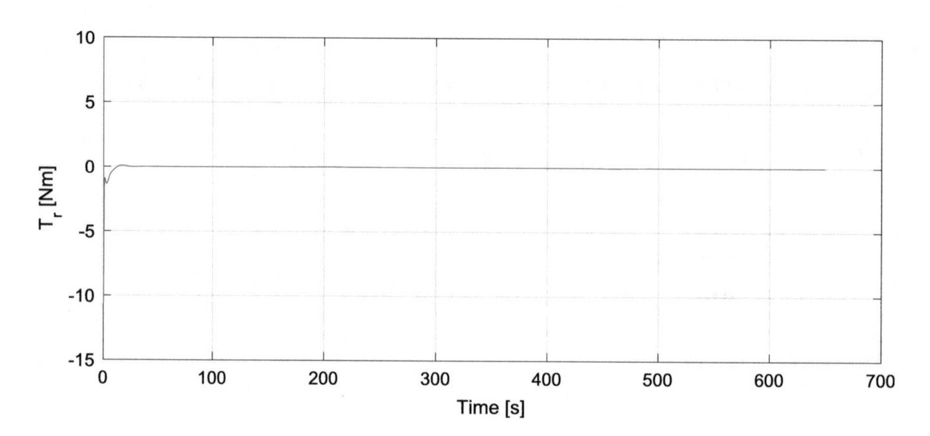

Fig. 5.7 Applied control action $\boldsymbol{\tau}_r$ versus time

5.2 Application to Aircraft

In this section, the trajectory tracking problem for the planar vertical take-off and landing (PVTOL) aircraft is analyzed. The system is nonlinear and multivariable. The model for this aircraft can be found in Gandolfo, Rosales, Patiño, Scaglia, & Jordan (2014), being summarized in (5.29), and represented in Fig. 5.8.

$$\begin{aligned}\ddot{x} &= -u_1 \sin\theta + \varepsilon \cos\theta u_2 \\ \ddot{y} &= u_1 \cos\theta + \varepsilon \sin\theta u_2 - 1 \\ \ddot{\theta} &= u_2\end{aligned} \tag{5.29}$$

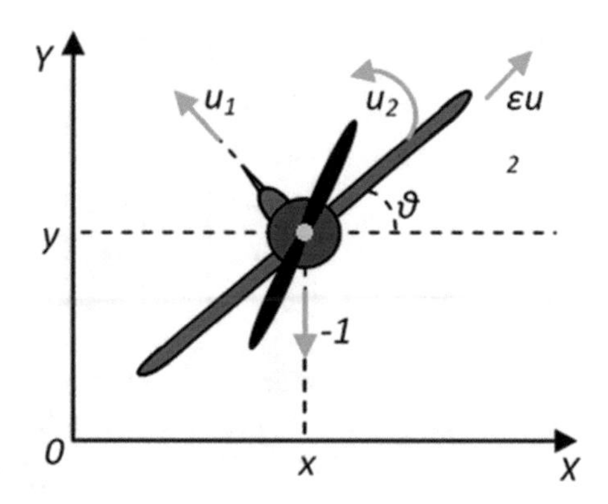

Fig. 5.8 Front view of PVTOL aircraft

The aircraft states are the position of the aircraft center of mass x, y, the roll angle θ of the aircraft, and the corresponding velocities $\dot{x}, \dot{y}, \dot{\theta}$. The control inputs u_1 and u_2 are the thrust (directed out the bottom of the aircraft) and the rolling moment about the aircraft center of mass, respectively. The parameter ε is a small coefficient, representing the coupling between the rolling moment and the lateral force. This parameter is usually negligible and uncertain, being assumed that $\varepsilon = 0$ (Fantoni, Lozano, & Palomino, 2003; Gandolfo et al., 2014; Hauser, Sastry, & Meyer, 1992). Thus, the simplified model is given by

$$\begin{aligned}\ddot{x} &= -u_1 \sin\theta \\ \ddot{y} &= u_1 \cos\theta - 1 \\ \ddot{\theta} &= u_2\end{aligned} \tag{5.30}$$

Let us now apply the LAB CD procedure as described in Chap. 2.

Step 1. Get an internal representation of the plant

$$\begin{aligned} x_1 &= x \rightarrow \dot{x}_1 = x_2 \\ x_2 &= \dot{x} \rightarrow \dot{x}_2 = -u_1 \sin x_5 \\ x_3 &= y \rightarrow \dot{x}_3 = x_4 \\ x_4 &= \dot{y} \rightarrow \dot{x}_4 = u_1 \cos x_5 - 1 \\ x_5 &= \theta \rightarrow \dot{x}_5 = x_6 \\ x_6 &= \dot{\theta} \rightarrow \dot{x}_6 = u_2 \end{aligned} \tag{5.31}$$

and discretize it

$$\begin{bmatrix} x_{1,n+1} \\ x_{2,n+1} \\ x_{3,n+1} \\ x_{4,n+1} \\ x_{5,n+1} \\ x_{6,n+1} \end{bmatrix} = \begin{bmatrix} x_{1,n} \\ x_{2,n} \\ x_{3,n} \\ x_{4,n} \\ x_{5,n} \\ x_{6,n} \end{bmatrix} + T \begin{bmatrix} x_{2,n} \\ -u_{1,n} \sin x_{5,n} \\ x_{4,n} \\ u_{1,n} \cos x_{5,n} - 1 \\ x_{6,n} \\ u_{2,n} \end{bmatrix} \tag{5.32}$$

Step 2. Split the state vector into two subvectors, collecting the state variables to be tracked, $(x, y) = (x_1, x_3)$, whose reference is given, and the remaining state variables, (x_2, x_4, x_5, x_6), denoted as sacrificed variables, whose reference will be determined, if so required.

Step 3. Define the next value of the state variables as an approximation to those of their references. This approximation is the control design stage.

Step 4. Combining (2.27) and (2.29), a model of the controlled plant (2.30) is obtained as

$$\underbrace{\begin{bmatrix} 0 & 0 \\ -\sin x_{5,\text{ref},n} & 0 \\ 0 & 0 \\ \cos x_{5,\text{ref},n} & 0 \\ 0 & 0 \\ 0 & 1 \end{bmatrix}}_{\mathbf{A}} \underbrace{\begin{bmatrix} u_{1,n} \\ u_{2,n} \end{bmatrix}}_{\mathbf{u}} = \underbrace{\begin{bmatrix} \frac{x_{1,\text{ref},n+1} - k_1(x_{1,\text{ref},n} - x_{1,n}) - x_{1,n}}{T} - x_{2,\text{ref},n} \\ \frac{x_{2,\text{ref},n+1} - k_2(x_{2,\text{ref},n} - x_{2,n}) - x_{2,n}}{T} \\ \frac{x_{3,\text{ref},n+1} - k_3(x_{3,\text{ref},n} - x_{3,n}) - x_{3,n}}{T} - x_{4,\text{ref},n} \\ \frac{x_{4,\text{ref},n+1} - k_4(x_{4,\text{ref},n} - x_{4,n}) - x_{4,n}}{T} + 1 \\ \frac{x_{5,\text{ref},n+1} - k_5(x_{5,\text{ref},n} - x_{5,n}) - x_{5,n}}{T} - x_{6,\text{ref},n} \\ \frac{x_{6,\text{ref},n+1} - k_6(x_{6,\text{ref},n} - x_{6,n}) - x_{6,n}}{T} \end{bmatrix}}_{\mathbf{b}} \tag{5.33}$$

Equation (5.33) represents a system of six linear equations with two unknown variables (u_1, u_2), and it allows the computation of the control actions at each sampling time instant so that the aircraft achieves the desired trajectory. Note that, if the vector on the right side of (5.33) is denoted as $\boldsymbol{b}$ and the vector that contains the control actions as $\boldsymbol{u}$, the system can be written in the classic equation system form $\boldsymbol{Au} = \boldsymbol{b}$.

Now, the condition for this system to have an exact solution is that vector $\boldsymbol{b}$ belongs to the column space of $\boldsymbol{A}$. Thus, from the first and third rows of system (5.33), it is possible to determine the value of the sacrificed variables $x_{2,\text{ref}}$ and $x_{4,\text{ref}}$,

$$x_{2,\text{ref},n} = \frac{x_{1,\text{ref},n+1} - k_1(x_{1,\text{ref},n} - x_{1,n}) - x_{1,n}}{T} \tag{5.34}$$

$$x_{4,\text{ref},n} = \frac{x_{3,\text{ref},n+1} - k_3(x_{3,\text{ref},n} - x_{3,n}) - x_{3,n}}{T} \tag{5.35}$$

Then, from the rows two and four, the necessary value of the orientation (x_5) to ensure the convergence to zero of the tracking errors can be obtained. Thus, from the second and fourth equations in (5.33)

$$\underbrace{\begin{bmatrix} -\sin x_{5,\text{ref},n} \\ \cos x_{5,\text{ref},n} \end{bmatrix}}_{A} u_{1,n} = \underbrace{\begin{bmatrix} \dfrac{x_{2,\text{ref},n+1} - k_2(x_{2,\text{ref},n} - x_{2,n}) - x_{2,n}}{T} \\ \dfrac{x_{4,\text{ref},n+1} - k_4(x_{4,\text{ref},n} - x_{4,n}) - x_{4,n}}{T} + 1 \end{bmatrix}}_{b} \tag{5.36}$$

It is easy to realize that (5.36) is a system of two equations with just one unknown variable, x_5. Thus, in order to have a unique solution, the vectors A and b should be parallel, that is,

$$-\tan x_{5,\text{ref},n} = -\frac{\sin x_{5,\text{ref},n}}{\cos x_{5,\text{ref},n}} = \frac{x_{2,\text{ref},n+1} - k_2(x_{2,\text{ref},n} - x_{2,n}) - x_{2,n}}{x_{4,\text{ref},n+1} - k_4(x_{4,\text{ref},n} - x_{4,n}) - x_{4,n} + T} \tag{5.37}$$

This orientation variable is defined as the reference for this sacrificed variable. Thus, $x_{5,\text{ref}}$ is computed as:

$$x_{5,\text{ref},n} = \text{atan}\left(-\frac{x_{2,\text{ref},n+1} - k_2(x_{2,\text{ref},n} - x_{2,n}) - x_{2,n}}{x_{4,\text{ref},n+1} - k_4(x_{4,\text{ref},n} - x_{4,n}) - x_{4,n} + T}\right) \tag{5.38}$$

The last condition required to ensure an exact solution of system (5.33) is determined by inspection of the fifth row:

$$x_{6,\text{ref},n} = \frac{x_{5,\text{ref},n+1} - k_5(x_{5,\text{ref},n} - x_{5,n}) - x_{5,n}}{T} \tag{5.39}$$

The conditions to get a unique solution of (5.33) have been obtained. That means, the desired trajectory ($x_{1,\text{ref}}$, $x_{3,\text{ref}}$) is an admissible trajectory, fully satisfying the plant model.

The control actions are summarized as

$$\underbrace{\begin{bmatrix} -\sin x_{5,\text{ref},n} & 0 \\ \cos x_{5,\text{ref},n} & 0 \\ 0 & 1 \end{bmatrix}}_{A} \underbrace{\begin{bmatrix} u_{1,n} \\ u_{2,n} \end{bmatrix}}_{u} = \underbrace{\begin{bmatrix} \dfrac{x_{2,\text{ref},n+1} - k_2(x_{2,\text{ref},n} - x_{2,n}) - x_{2,n}}{T} \\ \dfrac{x_{4,\text{ref},n+1} - k_4(x_{4,\text{ref},n} - x_{4,n}) - x_{4,n}}{T} + 1 \\ \dfrac{x_{6,\text{ref},n+1} - k_6(x_{6,\text{ref},n} - x_{6,n}) - x_{6,n}}{T} \end{bmatrix}}_{b} \tag{5.40}$$

And, finally, the control actions are obtained from (5.40) applying least square:

$$\begin{bmatrix} u_{1,n} \\ u_{2,n} \end{bmatrix} = \begin{bmatrix} -\boldsymbol{b}(1) \sin x_{5,\text{ref},n} + \boldsymbol{b}(2) \cos x_{5,\text{ref},n} \\ \boldsymbol{b}(3) \end{bmatrix} \tag{5.41}$$

Note that the resulting control law has no complicated terms to compute because most of them are simple operations that can be easily implemented in a simple microcontroller.

5.2.1 *Controller Performance*

Once more, let us analyze the tracking errors of the controlled plant.

The desired references for the sacrificed variables (x_2, x_4, x_5, x_6) are computed such that the following equation is satisfied

$$\begin{bmatrix} x_{1,\text{ref},n+1} - k_1(x_{1,\text{ref},n} - x_{1,n}) \\ x_{2,\text{ref},n+1} - k_2(x_{2,\text{ref},n} - x_{2,n}) \\ x_{3,\text{ref},n+1} - k_3(x_{3,\text{ref},n} - x_{3,n}) \\ x_{4,\text{ref},n+1} - k_4(x_{4,\text{ref},n} - x_{4,n}) \\ x_{5,\text{ref},n+1} - k_5(x_{5,\text{ref},n} - x_{5,n}) \\ x_{6,\text{ez},n+1} - k_6(x_{6,\text{ez},n} - x_{6,n}) \end{bmatrix} = \begin{bmatrix} x_{1,n} \\ x_{2,n} \\ x_{3,n} \\ x_{4,n} \\ x_{5,n} \\ x_{6,n} \end{bmatrix} + T \begin{bmatrix} x_{2,\text{ref},n} \\ -u_{1,n} \sin x_{5,\text{ref},n} \\ x_{4,\text{ref},n} \\ u_{1,n} \cos x_{5,\text{ref},n} - 1 \\ x_{6,\text{ref},n} \\ u_{2,n} \end{bmatrix} \tag{5.42}$$

Subtracting (5.42) and (5.32)

$$\begin{bmatrix} x_{1,\text{ref},n+1} - k_1(x_{1,\text{ref},n} - x_{1,n}) - x_{1,n+1} \\ x_{2,\text{ref},n+1} - k_2(x_{2,\text{ref},n} - x_{2,n}) - x_{2,n+1} \\ x_{3,\text{ref},n+1} - k_3(x_{3,\text{ref},n} - x_{3,n}) - x_{3,n+1} \\ x_{4,\text{ref},n+1} - k_4(x_{4,\text{ref},n} - x_{4,n}) - x_{4,n+1} \\ x_{5,\text{ref},n+1} - k_5(x_{5,\text{ref},n} - x_{5,n}) - x_{5,n+1} \\ x_{6,\text{ref},n+1} - k_6(x_{6,\text{ref},n} - x_{6,n}) - x_{6,n+1} \end{bmatrix} = T \begin{bmatrix} x_{2,\text{ref},n} - x_{2,n} \\ -u_{1,n}(\sin x_{5,\text{ref},n} - \sin x_{5,n}) \\ x_{4,\text{ref},n} - x_{4,n} \\ u_{1,n}(\cos x_{5,\text{ref},n} - \cos x_{5,n}) \\ x_{6,\text{ref},n} - x_{6,n} \\ 0 \end{bmatrix} \tag{5.43}$$

$$\begin{bmatrix} e_{1,n+1} - k_1 e_{1,n} \\ e_{2,n+1} - k_2 e_{2,n} \\ e_{3,n+1} - k_3 e_{3,n} \\ e_{4,n+1} - k_4 e_{4,n} \\ e_{5,n+1} - k_5 e_{5,n} \\ e_{6,n+1} - k_6 e_{6,n} \end{bmatrix} = T \begin{bmatrix} e_{2,n} \\ -u_{1,n}(\sin x_{5,\mathrm{ref},n} - \sin x_{5,n}) \\ e_{4,n} \\ u_{1,n}(\cos x_{5,\mathrm{ref},n} - \cos x_{5,n}) \\ e_{6,n} \\ 0 \end{bmatrix} \tag{5.44}$$

Again, by applying the Taylor series expansion to $\cos x_{5,n}$ and $\sin x_{5,n}$

$$\begin{bmatrix} e_{1,n+1} - k_1 e_{1,n} \\ e_{2,n+1} - k_2 e_{2,n} \\ e_{3,n+1} - k_3 e_{3,n} \\ e_{4,n+1} - k_4 e_{4,n} \\ e_{5,n+1} - k_5 e_{5,n} \\ e_{6,n+1} - k_6 e_{6,n} \end{bmatrix} = T \begin{bmatrix} e_{2,n} \\ u_{1,n} \sin\left(x_{5,\mathrm{ref},n} + \lambda_1 (x_{5,n} - x_{5,\mathrm{ref},n})\right) e_{5,n} \\ e_{4,n} \\ u_{1,n} \cos\left(x_{5,\mathrm{ref},n} + \lambda_2 (x_{5,n} - x_{5,\mathrm{ref},n})\right) e_{5,n} \\ e_{6,n} \\ 0 \end{bmatrix} \tag{5.45}$$

Looking at system (5.45), it can be seen that if $0 < k_6 < 1$, then $e_{6,n} \to 0$. Then, if $0 < k_5 < 1$, $e_{5,n} \to 0$ when $n \to \infty$. In the same way, if $0 < k_2, k_4 < 1$, $e_{4,n} \to 0$ and $e_{2,n} \to 0$ and $0 < k_1, k_3 < 1$, it results that $e_{1,n} \to 0$ and $e_{3,n} \to 0$, when $n \to \infty$.

5.2.2 Simulation Results

To illustrate the performance of the developed controller, a simulation was carried out in Simulink platform from MatLab software. The control structure was implemented according to Fig. 5.9. For the simulation, a sinusoidal reference profile for the aircraft and a null initial condition of all the states have been assumed. The used sampling period is $T = 0.1$ s, and the controller parameters are the following: $k_1 = 0.96$, $k_2 = 0.7$, $k_3 = 0.96$, $k_4 = 0.8$, $k_5 = 0.7$, and $k_6 = 0.7$.

The results of the test are shown in Fig. 5.10. As it can be seen, the aircraft follows the reference trajectory with a small error. The applied control actions, $\boldsymbol{u}_1$ and $\boldsymbol{u}_2$, are shown in Figs. 5.11 and 5.12, respectively. The aircraft planar position, angular position, and angular velocity are shown in Fig. 5.13a–d. By inspection of the presented results, it can be concluded that plant with the LAB controller presents good performance.

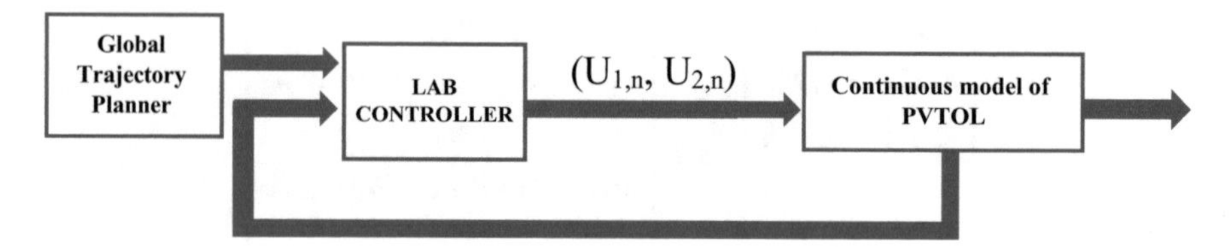

Fig. 5.9 Structure of the control system implementation

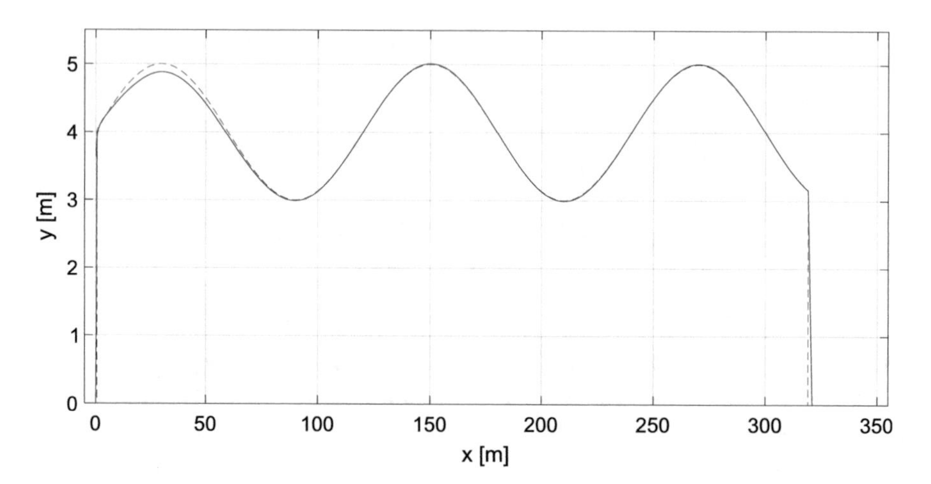

Fig. 5.10 Reference position (red dot line) and aircraft position (continuous blue line)

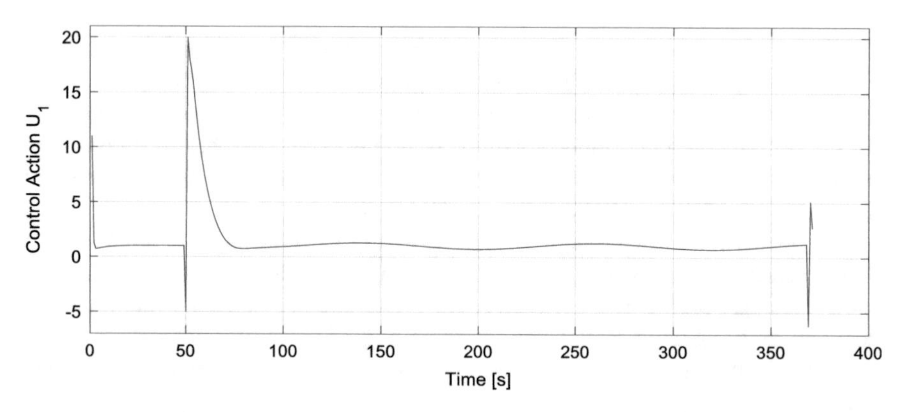

Fig. 5.11 Control action u_1 versus time

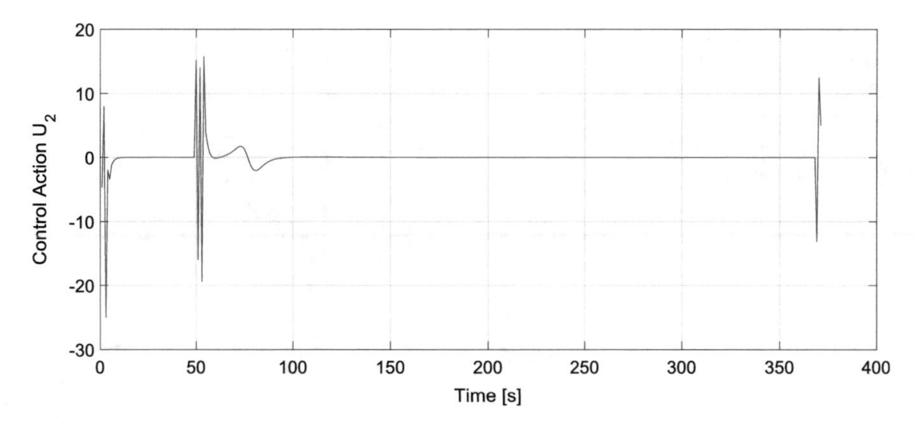

Fig. 5.12 Control action $\boldsymbol{u_2}$ versus time

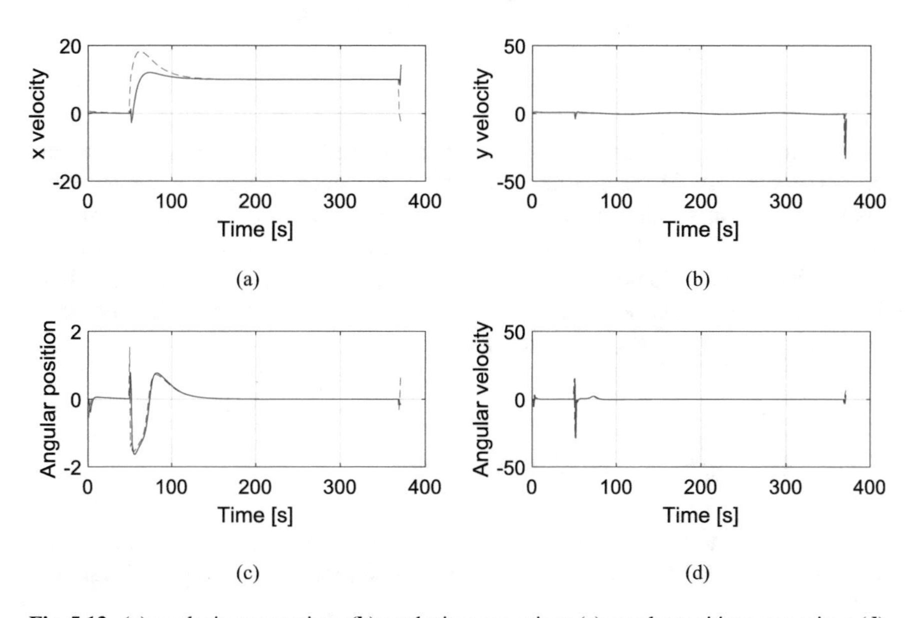

Fig. 5.13 (**a**) $\boldsymbol{x}$ velocity versus time; (**b**) $\boldsymbol{y}$ velocity versus time; (**c**) angular position versus time; (**d**) angular velocity versus time. Reference value: red dot line; variable value: continuous blue line

5.3 Quad Rotor Application

In the past decades, the research effort related to unmanned aerial vehicles (UAVs) has grown substantially, aiming at either military or civil applications. The use of an UAV is extremely advantageous in several tasks, compared to the use of unmanned ground vehicles (UGV), due to its tridimensional mobility. Unmanned aerial vehicles can be classified as fixed-wing, rotating-wing, and blimps. The main advantage

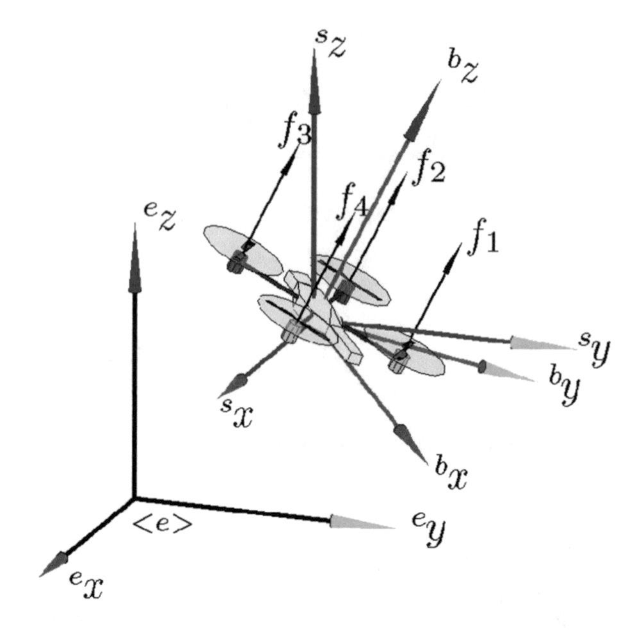

Fig. 5.14 Diagram of a 6-DOF quadrotor and the associated frames where e_i, s_i, and b_i represent the inertial, spatial, and body frame, respectively

of rotating-wing over fixed-wing aircraft is the ability of hovering and having omnidirectional movement. A disadvantage is, however, a relatively higher power consumption during the flight. Inside the rotating-wing aircraft classification, a quadrotor is much simpler and easier to build in comparison to a classical helicopter, since it has no swash plate and is controlled by only varying the angular velocity of the four motors. In this section, the LAB CD method is used to deal with trajectory tracking control problems in a quadrotor.

5.3.1 Dynamic Model of a Quadrotor

Consider the nonlinear model of a quadrotor, Fig. 5.14, given by (5.46) (Rosales, Gandolfo, Scaglia, Jordan, & Carelli, 2015; Salazar-Cruz, Palomino, & Lozano, 2005).

$$\begin{bmatrix} m\ddot{x} \\ m\ddot{y} \\ m\ddot{z} \\ \ddot{\phi} \\ \ddot{\theta} \\ \ddot{\psi} \end{bmatrix} = \begin{bmatrix} -u\sin\theta \\ u\cos\theta\sin\phi \\ u\cos\theta\cos\phi - mg \\ \tau_\phi \\ \tau_\theta \\ \tau_\psi \end{bmatrix} \tag{5.46}$$

where $\xi = (x, y, z)$ denotes the position of center of mass of the helicopter related to the inertial frame inertial frame, <*e*>, and $\eta = (\phi, \theta, \psi) \in R^3$ are the Euler angles. ϕ is the roll angle, θ is the pitch angle and ψ is the yaw angle in the spatial frame <*s*>. They represent the helicopter orientation (see Fig. 5.14). Where u, τ_ϕ, τ_θ, and τ_ψ are the impulse of the four propellers, the roll, the pitch and the yaw torque, respectively. The control objective is to find the combined control actions u, τ_ϕ, τ_θ and τ_ψ so that the UAV can reach and follow a reference trajectory($x_{\text{ref}}, y_{\text{ref}}, z_{\text{ref}}, \phi_{\text{ref}}$).

Now, the control design steps as described in Chap. 2 are applied to control the trajectory of the quadrotor (5.46).

First, an internal representation of the plant is obtained:

$$\begin{bmatrix} \dot{x}_1 \\ \dot{x}_2 \\ \dot{x}_3 \\ \dot{x}_4 \\ \dot{x}_5 \\ \dot{x}_6 \\ \dot{x}_7 \\ \dot{x}_8 \\ \dot{x}_9 \\ \dot{x}_{10} \\ \dot{x}_{11} \\ \dot{x}_{12} \end{bmatrix} = \begin{bmatrix} x_2 \\ -{}^u\!/_m \sin x_9 \\ x_4 \\ {}^u\!/_m \cos x_9 \sin x_7 \\ x_6 \\ {}^u\!/_m \cos x_9 \cos x_7 - g \\ x_8 \\ \tau_\phi \\ \dot{x}_{10} \\ \tau_\theta \\ x_{12} \\ \tau_\psi \end{bmatrix} \text{ where } \begin{bmatrix} x_1 \\ x_2 \\ x_3 \\ x_4 \\ x_5 \\ x_6 \\ x_7 \\ x_8 \\ x_9 \\ x_{10} \\ x_{11} \\ x_{12} \end{bmatrix} = \begin{bmatrix} x \\ \dot{x} \\ y \\ \dot{y} \\ z \\ \dot{z} \\ \phi \\ \dot{\phi} \\ \theta \\ \dot{\theta} \\ \psi \\ \dot{\psi} \end{bmatrix} \tag{5.47}$$

Next, a discrete version of (5.47) is obtained by using a Euler approximation:

$$\underbrace{\begin{bmatrix} 0 & 0 & 0 & 0 \\ 0 & 0 & 0 & 0 \\ 0 & 0 & 0 & 0 \\ 0 & 0 & 0 & 0 \\ 0 & 0 & 0 & 0 \\ 0 & 0 & 0 & 0 \\ -{}^T\!/_m \sin x_{9,n} & 0 & 0 & 0 \\ {}^T\!/_m \cos x_{9,n} \sin x_{7,n} & 0 & 0 & 0 \\ {}^T\!/_m \cos x_{9,n} \cos x_{7,n} & 0 & 0 & 0 \\ 0 & T & 0 & 0 \\ 0 & 0 & T & 0 \\ 0 & 0 & 0 & T \end{bmatrix}}_{A} \underbrace{\begin{bmatrix} u_n \\ \tau_{\phi,n} \\ \tau_{\theta,n} \\ \tau_{\psi,n} \end{bmatrix}}_{u} = \underbrace{\begin{bmatrix} x_{1,n+1} - x_{1,n} - Tx_{2,n} \\ x_{3,n+1} - x_{3,n} - Tx_{4,n} \\ x_{5,n+1} - x_{5,n} - Tx_{6,n} \\ x_{7,n+1} - x_{7,n} - Tx_{8,n} \\ x_{9,n+1} - x_{9,n} - Tx_{10,n} \\ x_{11,n+1} - x_{11,n} - Tx_{12,n} \\ x_{2,n+1} - x_{2,n} \\ x_{4,n+1} - x_{14n} \\ x_{6,n+1} - x_{6,n} + gT \\ x_{8,n+1} - x_{8,n} \\ x_{10,n+1} - x_{10,n} \\ x_{12,n+1} - x_{1,2n} \end{bmatrix}}_{b} \tag{5.48}$$

Then, following the procedure summarized in Chap. 2, the control problem is posed as a linear algebra problem, where the control actions are the unknown to be calculated at each sample time:

$$\begin{bmatrix} x_{1,n+1} \\ x_{2,n+1} \\ x_{3,n+1} \\ x_{4,n+1} \\ x_{5,n+1} \\ x_{6,n+1} \\ x_{7,n+1} \\ x_{8,n+1} \\ x_{9,n+1} \\ x_{10,n+1} \\ x_{11,n+1} \\ x_{12,n+1} \end{bmatrix} = \begin{bmatrix} x_{1,n} + Tx_{2,n} \\ x_{2,n} - T{}^{u}\!/_{m} \sin x_{9,n} \\ x_{3,n} + Tx_{4,n} \\ x_{4,n} + T{}^{u}\!/_{m} \cos x_{9,n} \sin x_{7,n} \\ x_{5,n} + Tx_{6,n} \\ x_{6,n} + T{}^{u}\!/_{m} \cos x_{9,n} \cos x_{7,n} - g \\ x_{7,n} + Tx_{8,n} \\ x_{8,n} + T\tau_{\phi,n} \\ x_{9,n} + Tx_{10,n} \\ \tau_{\theta,n} \\ x_{11,n} + Tx_{12,n} \\ x_{12,n} + T\tau_{\psi,n} \end{bmatrix} \tag{5.49}$$

Now, the next value of the state variables in (5.49) is defined as an approximation of those of their references. This approximation is the control design stage. In addition, in order that system (5.49) as an exact solution, the first six rows of it must be equal to zero. Thus, as the reference trajectory is known, and assuming a proportional approach of the system to this reference, the reference for the variables (x_2, x_4, x_6, and x_{12}) is calculated as:

$$x_{2,\text{ref},n} = \frac{x_{1,\text{ref},n+1} - k_1(x_{1,\text{ref},n} - x_{1,n}) - x_{1,n}}{T} \tag{5.50}$$

$$x_{4,\text{ref},n} = \frac{x_{3,\text{ref},n+1} - k_3(x_{3,\text{ref},n} - x_{3,n}) - x_{3,n}}{T} \tag{5.51}$$

$$x_{6,\text{ref},n} = \frac{x_{5,\text{ref},n+1} - k_5(x_{5,\text{ref},n} - x_{5,n}) - x_{5,n}}{T} \tag{5.52}$$

$$x_{12,\text{ref},n} = \frac{x_{11,\text{ref},n+1} - k_{11}(x_{11,\text{ref},n} - x_{11,n}) - x_{11,n}}{T} \tag{5.53}$$

In a similar way, the values of the reference for the variables x_8 and x_{10} are computed:

$$x_{8,\text{ref},n} = \frac{x_{7,\text{ref},n+1} - k_7(x_{7,\text{ref},n} - x_{7,n}) - x_{7,n}}{T} \tag{5.54}$$

$$x_{10,\mathrm{ref},n} = \frac{x_{9,\mathrm{ref},n+1} - k_9(x_{9,\mathrm{ref},n} - x_{9,n}) - x_{9,n}}{T} \tag{5.55}$$

To compute $x_{8,\mathrm{ref}}$ and $x_{10,\mathrm{ref}}$ used in (5.54) and (5.55), $x_{7,\mathrm{ref}}$ and $x_{9,\mathrm{ref}}$ must be calculated. These values can be computed by searching the requiring conditions for the system (5.49) to have an exact solution:

$$\begin{bmatrix} -\sin x_{9,n} \\ \cos x_{9,n} \sin x_{7,n} \\ \cos x_{9,n} \cos x_{7,n} \end{bmatrix} u_n = \frac{m}{T} \begin{bmatrix} x_{2,\mathrm{ref},n+1} - k_2(x_{2,\mathrm{ref},n} - x_{2,n}) - x_{2,n} \\ x_{4,\mathrm{ref},n+1} - k_4(x_{4,\mathrm{ref},n} - x_{4,n}) - x_{4,n} \\ x_{6,\mathrm{ref},n+1} - k_2(x_{2,\mathrm{ref},n} - x_{2,n}) - x_{2,n} + gT \end{bmatrix} \tag{5.56}$$

In order to have an exact solution of the system (5.56), (5.57) and (5.58) must be fulfilled

$$\tan x_{7,\mathrm{ref},n} = \frac{\sin x_{7,\mathrm{ref},n}}{\cos x_{7,\mathrm{ref},n}} = \frac{x_{4,\mathrm{ref},n+1} - k_4(x_{4,\mathrm{ref},n} - x_{4,n}) - x_{4,n}}{x_{6,\mathrm{ref},n+1} - k_6(x_{6,\mathrm{ref},n} - x_{6,n}) - x_{6,n} + gT} \tag{5.57}$$

$$\tan x_{9,\mathrm{ref},n} = -\frac{x_{2,\mathrm{ref},n+1} - k_2(x_{2,\mathrm{ref},n} - x_{2,n}) - x_{2,n}}{x_{4,\mathrm{ref},n+1} - k_4(x_{4,\mathrm{ref},n} - x_{4,n}) - x_{4,n}} \tag{5.58}$$

Then, by applying least square to solve system (5.49), the control actions that must be applied at each sampling time are obtained:

$$u_n = \frac{m}{T} \times \{-\Delta_{x,2} \sin x_{9,\mathrm{ref},n} + \Delta_{x,4} \cos x_{9,\mathrm{ref},n} \sin x_{7,\mathrm{ref},n} + (\Delta_{x,6} + gT) \cos x_{9,\mathrm{ref},n} \cos x_{7,\mathrm{ref},n}\} \tag{5.59}$$

$$\tau_{\phi,n} = \frac{\Delta_{x,8}}{T} \tag{5.60}$$

$$\tau_{\theta,n} = \frac{\Delta_{x,10}}{T} \tag{5.61}$$

$$\tau_{\psi n} = \frac{\Delta_{x,12}}{T} \tag{5.62}$$

where

$$\Delta_{x,2} = x_{2,\mathrm{ref},n+1} - k_2(x_{2,\mathrm{ref},n} - x_{2,n}) - x_{2,n} \tag{5.63}$$

$$\Delta_{x,4} = x_{4,\mathrm{ref},n+1} - k_4(x_{4,\mathrm{ref},n} - x_{4,n}) - x_{4,n} \tag{5.64}$$

$$\Delta_{x,6} = x_{6,\mathrm{ref},n+1} - k_6(x_{6,\mathrm{ref},n} - x_{6,n}) - x_{6,n} \tag{5.65}$$

$$\Delta_{x,8} = x_{8,\mathrm{ref},n+1} - k_8(x_{8,\mathrm{ref},n} - x_{8,n}) - x_{8,n} \tag{5.66}$$

$$\Delta_{x,10} = x_{10,\text{ref},n+1} - k_{10}(x_{10,\text{ref},n} - x_{10,n}) - x_{10,n} \tag{5.67}$$

$$\Delta_{x,12} = x_{12,\text{ref},n+1} - k_{12}(x_{12,\text{ref},n} - x_{12,n}) - x_{12,n} \tag{5.68}$$

5.3.2 *Controller Performance*

To analyze the controller performance, first, $e_{i,\ n} = (x_{i,\ \text{ref},\ n} - x_{i,\ n})$, $i \in \{1, 2 \cdots 12\}$ and $0 \leq k_i < 1$, $i \in \{1, 2 \cdots 12\}$ are defined.

Next, equations (5.60), (5.61), and (5.62) are replaced in (5.48), and (5.69) is given:

$$\begin{bmatrix} e_{8,n+1} \\ e_{10,n+1} \\ e_{12,n+1} \end{bmatrix} = \begin{bmatrix} k_8 & 0 & 0 \\ 0 & k_{10} & 0 \\ 0 & 0 & k_{12} \end{bmatrix} \begin{bmatrix} e_{8,n} \\ e_{10,n} \\ e_{12,n} \end{bmatrix} \Rightarrow \lim_{n\to\infty} \begin{bmatrix} e_{8,n} \\ e_{10,n} \\ e_{12,n} \end{bmatrix} = \begin{bmatrix} 0 \\ 0 \\ 0 \end{bmatrix} \tag{5.69}$$

From (5.48), it is clear that

$$\begin{bmatrix} x_{7,n+1} \\ x_{9,n+1} \\ x_{11,n+1} \end{bmatrix} = \begin{bmatrix} x_{7,n} \\ x_{9,n} \\ x_{11,n} \end{bmatrix} + T \begin{bmatrix} x_{8,n} \\ x_{10,n} \\ x_{12,n} \end{bmatrix} \tag{5.70}$$

that is,

$$\begin{bmatrix} x_{7,n+1} \\ x_{9,n+1} \\ x_{11,n+1} \end{bmatrix} = \begin{bmatrix} x_{7,n} \\ x_{9,n} \\ x_{11,n} \end{bmatrix} + T \begin{bmatrix} x_{8,\text{ref},n} - e_{8,n} \\ x_{10,\text{ref},n} - e_{10,n} \\ x_{12,\text{ref},n} - e_{12,n} \end{bmatrix} \tag{5.71}$$

Thus, by replacing (5.53), (5.54), and (5.55) in (5.71), it yields

$$\begin{bmatrix} e_{7,n+1} \\ e_{9,n+1} \\ e_{11,n+1} \end{bmatrix} = \begin{bmatrix} k_7 & 0 & 0 \\ 0 & k_9 & 0 \\ 0 & 0 & k_{11} \end{bmatrix} \begin{bmatrix} e_{7,n} \\ e_{9,n} \\ e_{11,n} \end{bmatrix} + T \begin{bmatrix} e_{8,n} \\ e_{10,n} \\ e_{12,n} \end{bmatrix} \tag{5.72}$$

Considering (5.70), the steady-state errors in (5.72) result

$$\lim_{n\to\infty}\begin{bmatrix} e_{7,n} \\ e_{9,n} \\ e_{11,n} \end{bmatrix} = \begin{bmatrix} 0 \\ 0 \\ 0 \end{bmatrix} \tag{5.73}$$

An analogous procedure is applied to (5.44), leading to

$$\begin{bmatrix} e_{2,n+1} \\ e_{4,n+1} \\ e_{6,n+1} \end{bmatrix} = \begin{bmatrix} k_2 & 0 & 0 \\ 0 & k_4 & 0 \\ 0 & 0 & k_6 \end{bmatrix} \begin{bmatrix} e_{2,n} \\ e_{4,n} \\ e_{6,n} \end{bmatrix} + \frac{T}{m} u_n \begin{bmatrix} -\sin x_{9,\mathrm{ref},n} + \sin x_{9,n} \\ \cos x_{9,\mathrm{ref},n} \sin x_{7,\mathrm{ref},n} - \cos x_{9,n} \sin x_{7,n} \\ \cos x_{9,\mathrm{ref},n} \cos x_{7,\mathrm{ref},n} - \cos x_{9,n} \cos x_{7,n} \end{bmatrix} \tag{5.74}$$

$$\lim_{n\to\infty}\begin{bmatrix} e_{2,n} \\ e_{4,n} \\ e_{6,n} \end{bmatrix} = \begin{bmatrix} 0 \\ 0 \\ 0 \end{bmatrix} \tag{5.75}$$

From (5.75), (5.50), (5.51), (5.52), and (5.73), it yields

$$\lim_{n\to\infty}\begin{bmatrix} e_{1,n} \\ e_{3,n} \\ e_{5,n} \\ e_{11,n} \end{bmatrix} = \begin{bmatrix} 0 \\ 0 \\ 0 \\ 0 \end{bmatrix} \tag{5.76}$$

5.3.3 Simulation Results

To evaluate the controller performance, a simulation is carried out in Simulink from MatLab. Thus, the continuous time model (5.47) is programed, and the controller developed is applied according to Fig. 5.15.

The values of the parameters of the quadrotor model and controller parameters are obtained from Rosales et al. (2015)

$$m = 0.5 \ \ \mathrm{kg} l = 0.24 \ \ \mathrm{m} g = 9.81 \ \ \mathrm{m/s^2}$$
$$I = \mathrm{diag}[3.83.87.1] * 10^{-3} \mathrm{Nms^2/rad}$$

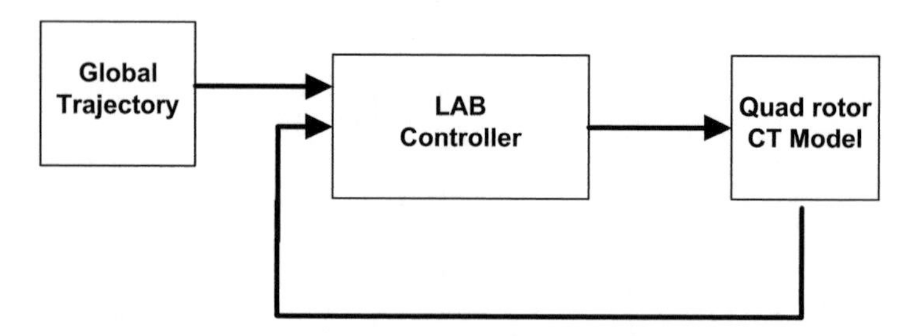

Fig. 5.15 Control architecture implemented for trajectory tracking in UAV

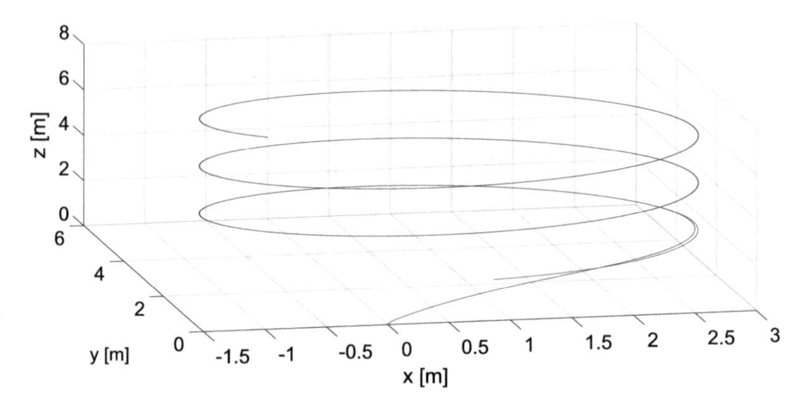

Fig. 5.16 Position of the quadrotor (blue line); reference trajectory (red line)

$$\begin{array}{llll} k_{x1} = 0.78 & k_{x2} = 0.8 & k_{x3} = 0.77 & k_{x4} = 0.8 \\ k_{x5} = 0.9 & k_{x6} = 0.95 & k_{x7} = 0.8 & k_{x8} = 0.85 \\ k_{x9} = 0.7 & k_{x10} = 0.67 & k_{x11} = 0.7 & k_{x12} = 0.67 \end{array}$$

A helical path of radius 2 m was used as the desired trajectory, centered at $(x, y) = (1\text{ m}, 3\text{ m})$. The initial position of reference path is $(x, y, z) = (1\text{ m}, 1\text{ m}, 1\text{ m})$, and the initial position of the quadrotor is at the system origin. The trajectory is generated with an upward velocity of $v_z = 0.05$ m/s and angular velocity of $\omega = 1$ rad/s.

In Fig. 5.16, the performance of the control system is presented. As it can be seen, the helicopter reaches and follows the desired trajectory, tending to the reference values. Furthermore, it can be noted that the quadrotor reaches the desired trajectory very quickly and then follows it. The time evolution of the coordinates x, y, z and ψ is shown in Fig. 5.16.

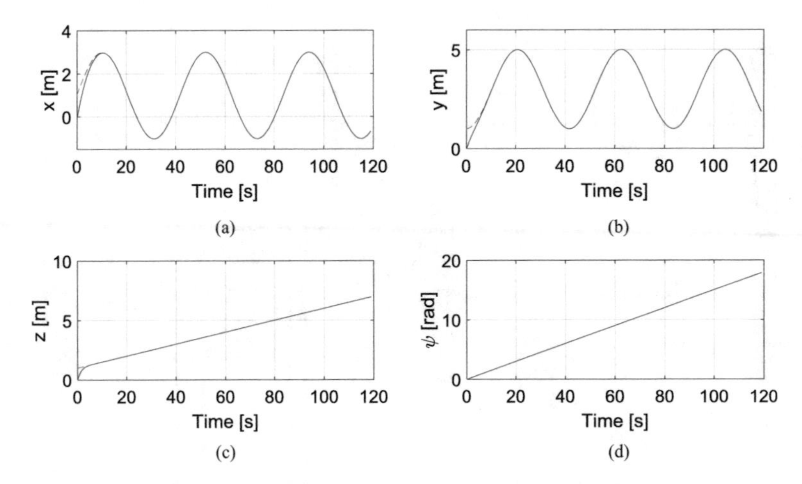

Fig. 5.17 (**a**) Red dot line: reference trajectory in x variable; blue line: x position of the UAV. (**b**) Red dot line: reference trajectory in y variable; blue line: y position of the quadrotor. (**c**) Red dot line: reference trajectory in z variable; blue line: z position of the UAV. (**d**) Red dot line: reference orientation ψ; blue line: current orientation position of the UAV

Appendix 5.1: Simulink Diagram for the Control of the Marine Vessel Described in Sect. 5.1

The implementation of the model on the Simulink platform is shown below. The general scheme of the implementation is shown in Fig. 5.18. As it can be seen, the connection of the blocks is carried out, following the LAB control structure. The subsystem vessel model, which is specified in Fig. 5.19, outputs the value of the state variables to the LAB controller block. The disturbance block is displayed in Fig. 5.20. This block allows the generation of the disturbances introduced into the ship model. The LAB controller block contains the embedded "controller.m" file, where the linear algebra methodology designed to follow a predefined trajectory is programed.

The controller is programed in LAB_CONTROLLER.m file:

```
function [yc]=controlador(entrada)
global tpc per1 per2 p m11c m22c m23c m32c m33c d11c d22c d23c d32c d33c
b11c b32c Ts xdeseado ydeseado titadeseado rr uu OMrr i vd kx ky kOM ku kr
k1 k2 OMref tu tr errx erry
global xan yan uan ran van OMan;

i=i+1;

Ts=0.1;
```

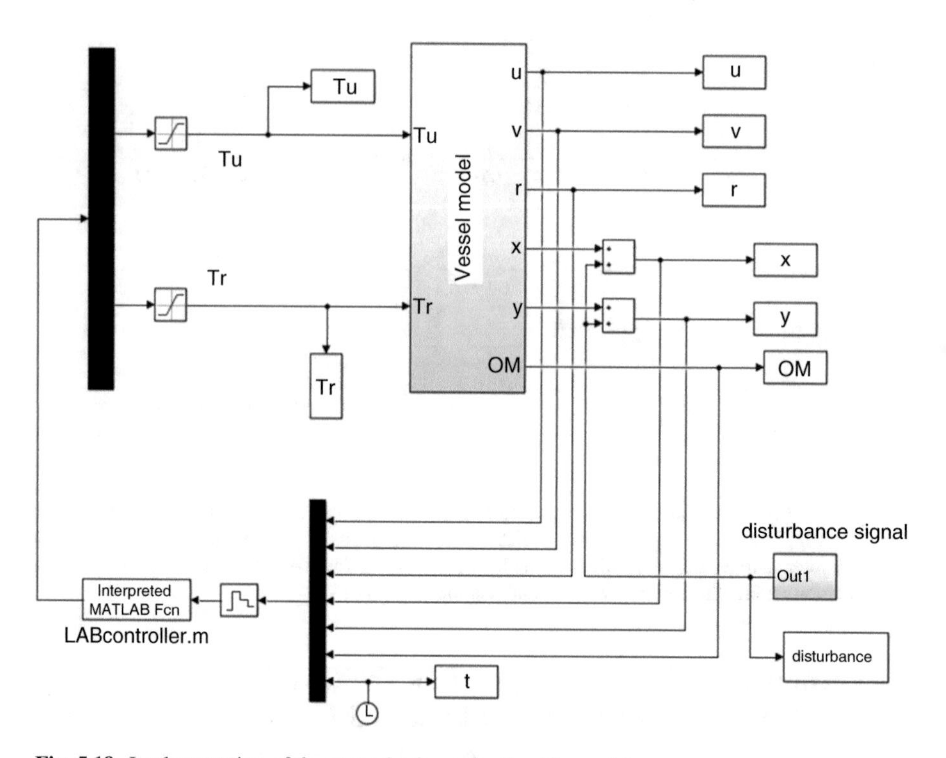

Fig. 5.18 Implementation of the general scheme for the ship model

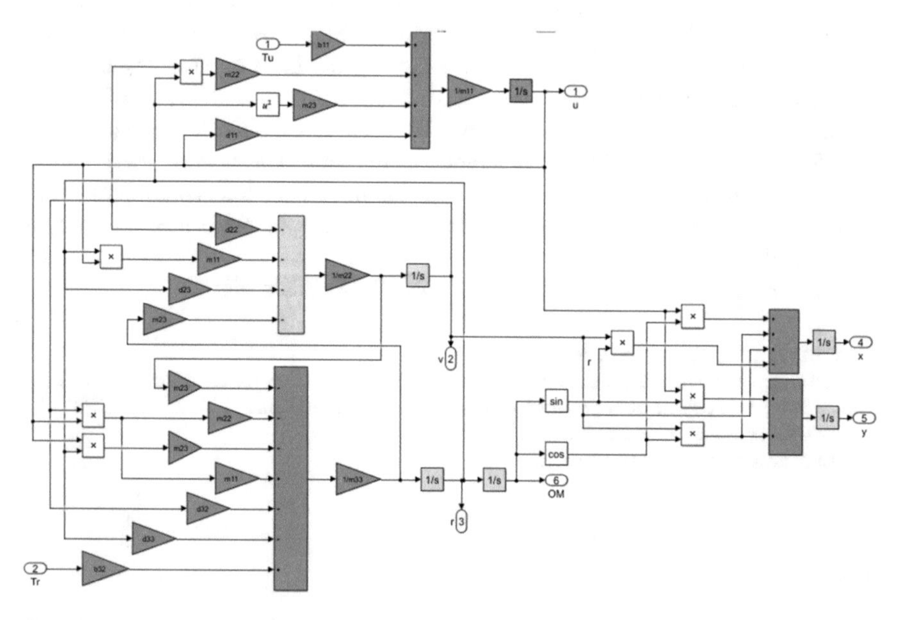

Fig. 5.19 Block connection for the implementation of the vessel model

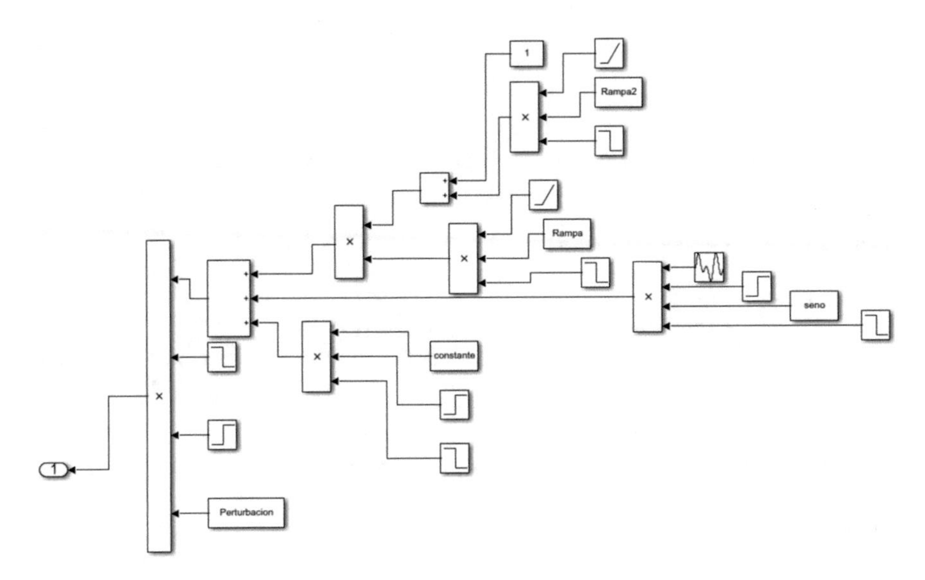

Fig. 5.20 Disturbance block subsystem implementation

```
ui =   entrada(1);
vi =   entrada(2);
ri =   entrada(3);
xi =   entrada(4);
yi =   entrada(5);
OMi =   entrada(6);
tiempo =   entrada(7);
OMrr(i)=OMi;
OMrr=unwrap(OMrr,pi);

%% -------------------
kinematic------------------------------------------

xd=   (xdeseado(i+1)  - kx*(xdeseado(i)-xi)  - xi)/Ts;
yd=   (ydeseado(i+1)  - ky*(ydeseado(i)-yi)  - yi)/Ts;
titadeseado(i)=atan2(ydeseado(i+1)- ydeseado(i), xdeseado(i+1)-
xdeseado(i));

aux = [xd + vi*sin(OMi) ; yd - vi*cos(OMi) ];
OMref(i+1)= atan2(aux(2) , aux(1))  ;
OMref=unwrap(OMref,pi);
OMref(i+1)=(OMref(i+1)+ (Tf1/Ts)*OMref(i))/(1+Tf1/Ts);
rr(i+1)= (OMref(i+1) - kOM*(OMref(i)-OMrr(i)) - OMrr(i)) / Ts;
uu(i+1)=pinv([cos(OMref(i+1));sin(OMref(i+1))])*aux;

udeseado=uu;
rdeseado=rr;
ud=(udeseado(i+1)  - ku*(udeseado(i)-ui)  - ui)/Ts;
rd=(rdeseado(i+1)  - kr*(rdeseado(i)-ri)  - ri)/Ts;
```

```
if i==0
  van=0;
end

vd(i)=-1/m22c*( m23c*rd + m11c*ui*ri + d22c*vi+ d23c*ri );

%% ------------------
dynamic------------------------------------------

A=[b11c 0;0 b32c];
f1= m11c*ud - m22c*vi*ri - m23c*ri^2 + d11c*ui;
f2= m23c*vd(i) +m33c*rd+m22c*vi*ui +m23c*ri*ui - m11c*vi*ui +d32c*vi +
d33c*ri;
b=[f1;f2];

yc=pinv(A)*b; %% ------------------ CONTROL ACTION

 end
```

References

Børhaug, E., Pavlov, A., Panteley, E., & Pettersen, K. Y. (2011). Straight line path following for formations of underactuated marine surface vessels. *IEEE Transactions on Control Systems Technology, 19*, 493–506.

Fantoni, I., Lozano, R., & Palomino, A. (2003). "Global stabilizing control design for the PVTOL aircraft using saturation functions on the inputs," Heudiasyc, UMR CNRS 6599, UTC, BP 20529, 60205 Compiegne, France. September 2013.

Gandolfo, D., Rosales, C., Patiño, D., Scaglia, G., & Jordan, M. (2014). Trajectory tracking control of a pvtol aircraft based on linear algebra theory. *Asian Journal of Control, 16*(6), 1849–1858.

Hauser, J., Sastry, S., & Meyer, G. (1992). Nonlinear control design for slightly nonminimum phase systems: Application to V/STOL aircraft. *Automatica, 28*(4), 665–679.

Rosales, C., Gandolfo, D., Scaglia, G., Jordan, M., & Carelli, R. (2015). Trajectory tracking of a mini four-rotor helicopter in dynamic environments-a linear algebra approach. *Robotica, 33*(8), 1628–1652.

Salazar-Cruz, S., Palomino, A., & Lozano, R. (2005, December). Trajectory tracking for a four rotor mini-aircraft. In *Proceedings of the 44th IEEE conference on decision and control* (pp. 2505–2510). New York: IEEE.

Serrano, M. E., Scaglia, G. J., Godoy, S. A., Mut, V., & Ortiz, O. A. (2013). Trajectory tracking of underactuated surface vessels: A linear algebra approach. *IEEE Transactions on Control Systems Technology, 22*(3), 1103–1111.

Chapter 6
Application to Industrial Processes

Most chemical industrial processes present a nonlinear behavior, leading to the development of nonlinear controllers to fulfill some requirements. If the main goal is to keep a reference, the regulation problem is usually simplified by considering a linearized model around the equilibrium point. If the main purpose of the control system is to follow a reference, for instance, to keep a process variable following a desired profile, the linearization is doubtful, and a more complicated controller is required.

In this chapter, this kind of process is considered, and Linear Algebra-Based Control Design will be the approach used to design the controller. First, the case of dealing with a nonlinear detailed model of the process will be considered. Then, an experimental linearized model will be considered and, again, LAB CD methodology will be applied to design the control.

The simple model based on a first-order plus time delay (FOPTD) transfer function will be considered. Thence, a design applicable to a large variety of processes is obtained. If the state evolution goes far from the equilibrium point, where the linearized model has been obtained, a battery of linear models can be obtained along the process trajectory, and a sort of gain-scheduling adaptation can be foreseen.

In order to better illustrate the procedure, a typical continuous stirred tank reactor (CSTR) will be considered.

6.1 Continuous Stirred Tank Reactor Model

In a simple CSTR, an exothermic reaction $A \rightarrow B$ takes place. The heat of reaction is removed in a cooling jacket which surrounds the reactor. The reaction is first order in reactant A, and negligible heat losses and perfectly mixing are assumed (Camacho & Smith, 2000; Coughanowr, Ansara, Luoma, Hamalainen, & Lukas, 1991; Luyben, 1990; Ray, 1981; Stephanopoulos & Vallino, 1991). A typical CSTR is depicted in

G. Scaglia et al., *Linear Algebra Based Controllers*,
https://doi.org/10.1007/978-3-030-42818-1_6

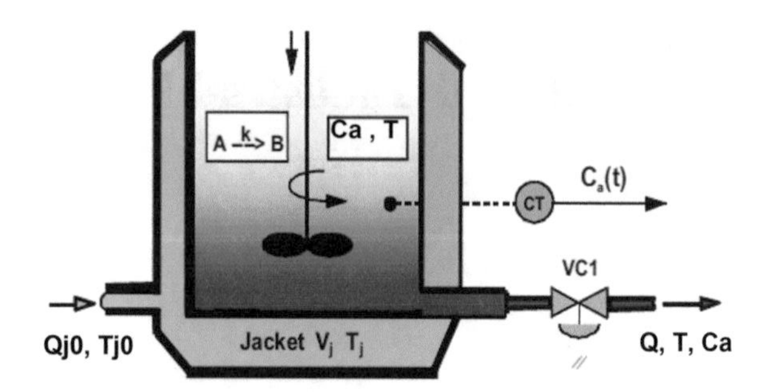

Fig. 6.1 CSTR

Fig. 6.1, where the total volume V is constant. The model reactor based on first principles is as follows:

– Component A mass balance

$$\frac{dC_a(t)}{dt} = \frac{q}{V}(C_{ao} - C_a(t)) - k(t)C_a{}^2(t); k(t) = k_o e^{\frac{-E}{R(T(t)+273)}} \tag{6.1}$$

where C_{ao} and C_a are the reactant concentrations for the inlet and outlet streams, respectively, q is the reactor flow, T is the reactor temperature, E is the activation energy, R is the perfect-gas constant, and k_o is the pre-exponential factor from Arrhenius law, assuming a reaction rate proportional to the square of the concentration. This exponential temperature dependence represents one of the most severe nonlinearities in dynamical systems.

– Energy balance for the reactor

$$\frac{dT(t)}{dt} = \frac{q}{V}(T_i - T(t)) - k(t)C_a{}^2(t)\frac{H}{\rho c_p} - \frac{UA}{\rho c_p V}\left(T(t) - T_j\right) \tag{6.2}$$

where H is the heat of reaction, ρ and c_p are the density and the heat capacity, respectively, for the inlet and outlet streams, U is the overall heat transfer coefficient in the jacket, A is the heat transfer area, T_j is the temperature of the refrigerated jacket, and T_i the inlet stream temperature.

– Energy balance for the jacket

$$\frac{dT_j}{dt} = \frac{UA}{\rho_j c_j V_j}\left(T(t) - T_j\right) + \frac{q_j(t)}{V_j}\left(T_{jo}(t) - T_j\right) \tag{6.3}$$

where T_j is the jacket temperature and T_{jo} is the inlet temperature of the refrigerator. $q_j(t)$ is the refrigerating flow, and it will be considered as the control input to the

CSTR. V_j is the volume of the jacket, and ρ_j and c_j are the density and the heat capacity refrigerator liquid, respectively.

The control goal is to manipulate the refrigerator flow, $q_j(t)$, to keep the concentration of the reactor output, following a predefined profile $C_{a,\ \mathrm{ref}}(t)$.

6.2 Linear Algebra-Based Control of the Continuous Stirred Tank Reactor

The procedure outlined in Chap. 2 will be followed to design the LAB control of the CSTR in DT.

Step 1 From (6.1)–(6.3) a simple DT model is derived

$$C_{a,n+1} = C_{a,n} + T\left(\frac{q}{V}(C_{ao} - C_{a,n}) - k_n C_{a,n}^2\right) \tag{6.4}$$

$$T_{n+1} = T_n + T\left(\frac{q}{V}(T_i - T_n) - k_n C_{a,n}^2 \frac{H}{\rho c_p} - \frac{UA}{\rho c_p V}(T_n - T_{j,n})\right) \tag{6.5}$$

$$T_{j,n+1} = T_{j,n} + T\left(\frac{UA}{\rho_j c_j V_j}(T_n - T_{j,n}) - \frac{q_{j,n}}{V_j}\cdot(T_{ji} - T_{j,n})\right) \tag{6.6}$$

Some of the defined constant variables, such as q, V, C_{a0}, T_i, and T_{ji}, could be time varying. In this case, they would be considered as disturbances.

Step 2 Rewrite the previous model to be inverted, assuming a proportional approaching for the state variables

$$\begin{bmatrix} 0 \\ 0 \\ -\dfrac{(T_{j,n} - T_{ji})}{V_j} \end{bmatrix} q_{j,n} = \begin{bmatrix} C_{a,\mathrm{ref},n+1} - k_{C_a}(C_{a,\mathrm{ref},n} - C_{a,n}) - C_{a,n} - T\left(\frac{q}{V}(C_{ao} - C_{a,n}) - k_n C_{a,n}^2\right) \\ T_{\mathrm{ref},n+1} - k_T(T_{\mathrm{ref},n} - T_n) - T_n - T\left(\frac{q}{V}(T_i - T_n) - k_n C_{a,n}^2 \frac{H}{\rho c_p} - \frac{UA}{\rho c_p V}(T_n - T_{j,n})\right) \\ \dfrac{T_{j,\mathrm{ref},n+1} - k_{T_j}(T_{j,\mathrm{ref},n} - T_{j,n}) - T_{j,n}}{T} - \dfrac{UA}{\rho_j c_j V_j}(T_n - T_{j,n}) \end{bmatrix} \tag{6.7}$$

Step 3 Assign the references of the sacrificed variables to get an exact solution of (6.7). Note that the first two rows should vanish. Then, denoting

$$k_{\mathrm{ref},n} = \left(-\frac{C_{a,\mathrm{ref},n+1} - k_{C_a}(C_{a,\mathrm{ref},n} - C_{a,n}) - C_{a,n}}{T} + \frac{q}{V}(C_{ao} - C_{a,n})\right)\frac{1}{C_{a,n}^2} \tag{6.8}$$

and taking into account (6.1), the reactor temperature reference should be

$$T_{\mathrm{ref},n} = \frac{-E/R}{\ln\left(k_{\mathrm{ref},n}/k_0\right)} - 273 \tag{6.9}$$

From the second row of (6.7), the reference for the jacket temperature should be

$$T_{j,\mathrm{ref},n} = \left(\frac{T_{\mathrm{ref},n+1} - k_T(T_{\mathrm{ref},n} - T_n) - T_n}{T} - \frac{q}{V}(T_i - T_n) + k_n C_{a,n}^2 \frac{H}{\rho c_p} + \frac{UA}{\rho c_p V} T_n\right) \times \frac{\rho c_p V}{UA} \tag{6.10}$$

Step 4 Compute the control action. From the third row in (6.7), it yields

$$q_{j,n} = \left(\frac{T_{j,\mathrm{ref},n+1} - k_{T_j}\left(T_{j,\mathrm{ref},n} - T_{j,n}\right) - T_{j,n}}{T} - \frac{UA}{\rho_j c_j V_j}\left(T_n - T_{j,n}\right)\right) \times \frac{V_j}{\left(T_{ji} - T_{j,n}\right)} \tag{6.11}$$

Note that in (6.11), the next value of the reference for the jacket temperature is required, but only its current value is computed by (6.10). So, based on the sequence $\{T_{j,\,\mathrm{ref},\,n}\}$, its next value should be estimated. The same happens with the temperature reference for the reactor appearing in (6.10) and being computed by (6.9).

6.3 Linear Algebra-Based Control Applied to a Simulated Continuous Stirred Tank Reactor

In order to evaluate the designed controller, a model of the reactor, as given in (6.4)–(6.6), with the parameters described in Table 6.1 has been simulated.

Remark 6.1 It is worth noting that the full process state should be available to compute (6.11) and (6.10).

In Fig. 6.2, the reactor concentration evolution for a sustained oscillation after a sudden change in its reference is plotted.

The evolution of the reactor temperature is shown in Fig. 6.3, and the temperature of the jacket is plotted in Fig. 6.4.

The control signal is plotted in Fig. 6.5.

Table 6.1 Nominal values and parameters of the reactor

Variable	Description	Value
C_{ao}	Reactive inlet concentration (kmol A/m^3)	2.88
V	Reactor volume (m^3)	7.08
T_i	Reactor flow inlet temperature (°C)	56
V_j	Jacket volume (m^3)	1.82
k_o	Pre exponential Arrhenius factor (m^3 s^{-1} kmol)	7.44×10^{-2}
E	Activation energy (J/kmol)	-1.182×10^7
U	Heat transmission coeff. J/(s m^2 °C)	3550
A	Heat transmission surface (m^2)	5.4
T_{ji}	Refrigerator flow inlet temperature (°C)	7
R	Gass constant J/kmol K	8314
H	Reaction heat (J/kmol)	-9.86×10^7
c_p	Thermal capacity of rector (J/kmol °C)	1.815×10^5
c_{pj}	Thermal capacity of jacket (J/kg °C)	4184
ρ	Reactive density (kmol/m^3)	19.2
ρ_j	Refrigerator density (kg/m^3)	1000
Operating point		
q	Reactor input flow (m^3/s)	7.5×10^{-3}
q_j	Maximum refrigerant input flow (m^3/s)	0.02

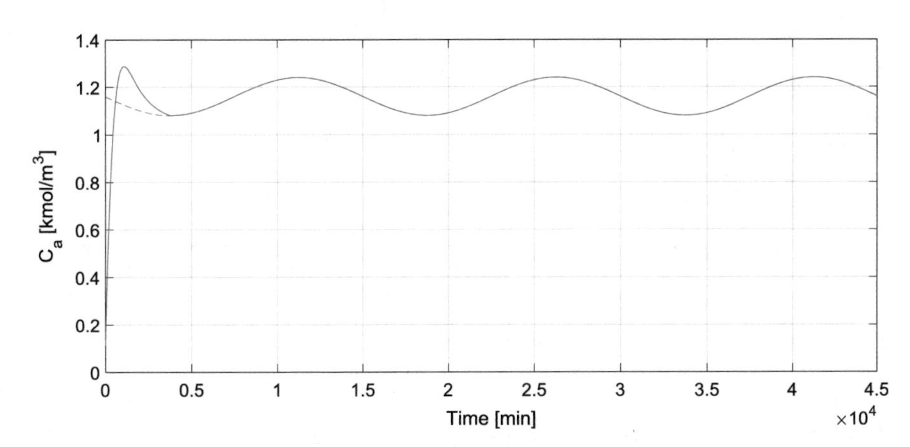

Fig. 6.2 Reference concentration (dashed line) and actual concentration

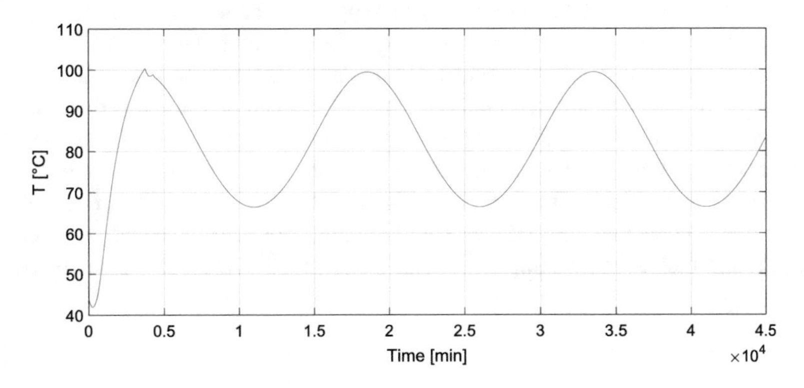

Fig. 6.3 Reactor temperature

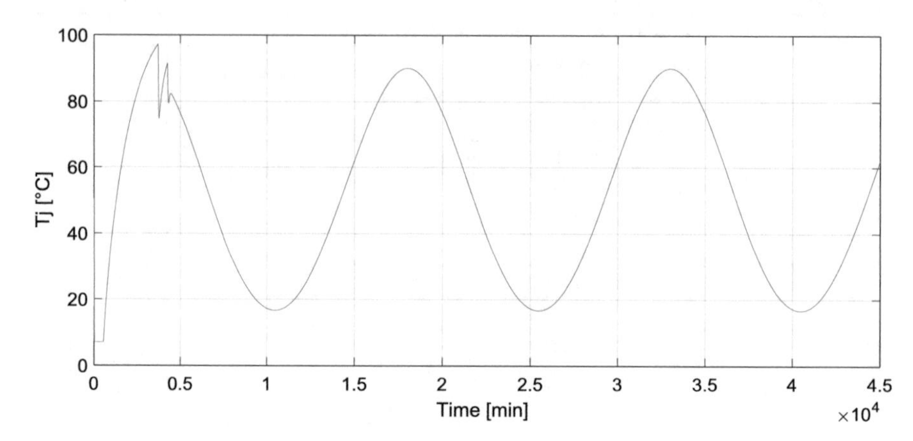

Fig. 6.4 Reactor jacket temperature

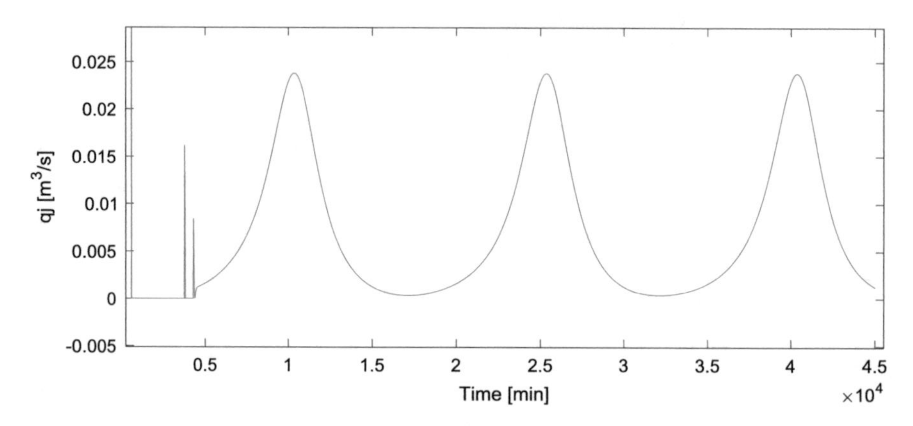

Fig. 6.5 Refrigerator flow evolution (control signal)

6.4 Linearized Model of Industrial Processes

As it is well known, a linearized model of the plant can be used if the control goal is to keep the process stabilized around an equilibrium point. The model (6.1)–(6.3) can be linearized but, due to the model parameters uncertainty as well as the missing of some additional dynamics, an experimental model can be used. For that purpose, first-order plus dead-time (FOPDT) models are easily calculated and provide good performance, comparable to higher-order models, whose simplicity and general utility make them highly practical (Liu, Wang, & Huang, 2013). Their main advantage is their ability to capture the essential dynamics of several industrial processes even though they are very simple (Seborg, Edgar, Mellichamp, & Doyle III, 2011).

One drawback of these reduced-order models is that they present uncertainties that lead to performance degradation of conventional controllers, such as proportional integral derivative (PID) or Smith Predictors (Camacho & Smith, 2000).

The general CT domain transfer function of an FOPDT model is given as:

$$G_1(s) = \frac{\Delta y(s)}{\Delta u(s)} = \frac{K}{\tau s + 1} \quad e^{-t_0 s} \tag{6.12}$$

where K is the steady-state process gain, τ is time constant, and t_o is the time delay or dead time. In order to deal with rational transfer functions, the dead time can be approximated by the first-order term of its Taylor series, rendering the following expression:

$$e^{-t_0 s} \approx \frac{1}{t_0 s + 1} \tag{6.13}$$

By replacing (6.13) in (6.12), the approximated expression of the transfer function takes the form:

$$G_2(s) \simeq \frac{K}{(\tau s + 1)\,(t_0 s + 1)} = \frac{K \; K_B}{s^2 + K_A s + K_B} \tag{6.14}$$

where

$$K_A = \frac{t_0 + \tau}{t_0 \;\; \tau} \qquad K_B = \frac{1}{t_0 \;\; \tau} \tag{6.15}$$

From (6.14), the following equation (6.16) can be written:

$$\ddot{y} + K_A \dot{y} + K_B (y - y_o) = K K_B (u - u_0) \tag{6.16}$$

where y_o and u_o denote the initial values of the measurable variable and the control action, respectively.

6.5 Linear Algebra Base Applied to a Linearized Model of Industrial Processes

Now the controller design can be performed, following the steps mentioned in Chap. 2. The goal is to find the values of the control action u, making the process to follow a reference trajectory with a minimum error (Sardella, Serrano, Camacho, & Scaglia, 2019).

The state variables and the input of the system can be defined as:

$$\begin{aligned} x_1 &= y - y_0 \\ x_2 &= \dot{y} \\ u_1 &= u - u_0 \end{aligned} \tag{6.17}$$

and the external representation (6.16) can be rewritten as

$$\begin{bmatrix} \dot{x}_1 \\ \dot{x}_2 \end{bmatrix} = \begin{bmatrix} 0 & 1 \\ -K_B & -K_A \end{bmatrix} \begin{bmatrix} x_1 \\ x_2 \end{bmatrix} + \begin{bmatrix} 0 \\ KK_B \end{bmatrix} \quad u_1 \tag{6.18}$$

Now, the discretization is performed by applying the Euler approximation to equation (6.18):

$$\begin{bmatrix} x_{1,n+1} \\ x_{2,n+1} \end{bmatrix} = \begin{bmatrix} x_{1,n} \\ x_{2,n} \end{bmatrix} + T \left\{ \begin{bmatrix} 0 & 1 \\ -K_B & -K_A \end{bmatrix} \begin{bmatrix} x_{1,n} \\ x_{2,n} \end{bmatrix} + \begin{bmatrix} 0 \\ KK_B \end{bmatrix} u_{1,n} \right\} \tag{6.19}$$

where T is the sampling period. Remember that the values of $x_1(t)$, $x_2(t)$, and u(t) at the discrete time $t = n\ T$, with $n\ \in \{0, 1, 2, \cdots\}$, are denoted as $x_{1,n}$, $x_{2,n}$, and u_n, respectively.

The tracking trajectory problem is now posed as a linear algebra problem, where the control action calculation can be done by solving the following linear equations system:

$$A\ u_{1,n} = b$$

$$\underbrace{\begin{bmatrix} 0 \\ KK_B \end{bmatrix}}_{A} u_{1,n} = \underbrace{\begin{bmatrix} x_{1,n+1} - x_{1,n} - Tx_{2,n} \\ \dfrac{x_{2,n+1} - x_{2,n}}{T} + K_A x_{2,n} + K_B x_{1,n} \end{bmatrix}}_{b} \tag{6.20}$$

To assure the solvability of system (6.20), the condition rank(A) = rank ([A b]) must be satisfied. This condition implies that $x_{2,n}$ must fulfill:

$$\begin{aligned} x_{1\text{ref},n+1} - k_1 e_{1,n} - x_{1,n} - T x_{2\text{ref},n} = 0 \quad \Rightarrow \\ x_{2\text{ref},n} = \frac{x_{1\text{ref},n+1} - k_1 e_{1,n} - x_{1,n}}{T} \end{aligned} \tag{6.21}$$

The value of $x_{2\text{ref},n}$ that satisfies (6.21) is called the reference of the sacrificed variable. The value adopted by such variable forces the equation system to have an exact solution and leads the tracking error to zero.

The tracking error e_2, defined in (6.23), should tend to zero. This condition can be written as follows:

$$e_{2,n+1} = k_2 \cdot e_{2,n} \tag{6.22}$$

$$\underbrace{x_{2\text{ref},\ n+1} - x_{2,\ n+1}}_{e_{2,n+1}} = k_2 \underbrace{(x_{2\text{ref},\ n} - x_{2,\ n})}_{e_{2,n}} \Rightarrow x_{2,n+1} = x_{2\text{ref},n+1} - k_2 e_{2,n} \tag{6.23}$$

Following an analogous procedure to the one applied for the tracking error on x_2, the expression for the error on x_1 can be written as:

$$e_{1,n+1} = k_1 \cdot e_{1,n} \tag{6.24}$$

$$\underbrace{x_{1\text{ref},\ n+1} - x_{1,\ n+1}}_{e_{1,n+1}} = k_1 \underbrace{(x_{1\text{ref},\ n} - x_{1,\ n})}_{e_{1,n}} \Rightarrow x_{1,n+1}$$
$$= x_{1\text{ref},n+1} - k_1 e_{1,n}, \quad 0 < k_1 < 1 \tag{6.25}$$

Finally, solving system (6.20) by least square, the control action law is obtained

$$u_{1,n} = \frac{1}{K K_B} \left[\frac{x_{2\text{ref},n+1} - k_2 e_{2,n} - x_{2,n}}{T} + K_A x_{2,n} + K_B x_{1,n} \right] \tag{6.26}$$

The design parameter k_i can take values between 0 and 1. Note that:

- if $k_i = 0$, $(y_{n+1} = y_{\text{ref},\ n+1})$, the goal is to reach the reference trajectory in one step.
- if $k_i = 1$, the error will remain constant $(e_{\text{i},\ n+1} = k_i \cdot e_{\text{i},\ n})$.

Remark 6.2 As always, the value of $x_{2,\ \text{ref},\ n+1}$ is required to compute $u_{1,\ n}$, u_n, but what (6.21) allows to calculate is $x_{2,\ \text{ref},\ n}$. However, $x_{2,\text{ref},n+1}$ can be estimated extrapolating the sequence. For instance, using the Taylor's formula:

$$x_{2\text{ref},n+1} = x_{2,\text{ref},n} + \frac{dx_{2\text{ref},n}}{dt} T + \frac{d^2 x_{2,\text{ref},n}}{dt^2} \frac{T^2}{2} + \ldots + C \tag{6.27}$$

where C is the complementary term. So, if the sampling time is small, $x_{2,\text{ref},\ n+1}$, can be approximated in one of the following ways:

$$x_{2,\text{ref},n+1} \approx x_{2,\text{ref},n} \tag{6.28}$$

$$x_{2,\text{ref},n+1} \approx x_{2,\text{ref},n} + \frac{dx_{2,\text{ref},n}}{dt} T \approx 2x_{2,\text{ref},n} - x_{2,\text{ref},n-1} \tag{6.29}$$

$$x_{2,\text{ref},n+1} \approx x_{2,\text{ref},n} + \frac{x_{2,\text{ref},n} - x_{2,\text{ref},n-1}}{T} T$$
$$+ \frac{x_{2,\text{ref},n} - 2x_{2,\text{ref},n-1} + x_{2,\text{ref},n-2}}{T^2} \frac{T^2}{2} \tag{6.30}$$

The first approximation, equation (6.28), provides excellent results for small sampling times. However, equation (6.29) was chosen to compute the control law because it gives lower deviations, being a simple expression.

Remark 6.3 The control law (6.26) involves a feedforward of the reference and a feedback from the state. But in this case, the state is the output and its derivative (6.17). Thus, a nonlinear PD controller weighting the output and its derivative can be used to implement the feedback term.

To evaluate the tracking errors' behavior, replacing $u_{1,n}$ from (6.26) in (6.19)

$$\begin{bmatrix} x_{1,n+1} \\ x_{2,n+1} \end{bmatrix} = \begin{bmatrix} x_{1,n} + T x_{2,n} \\ x_{2,\mathrm{ref},n+1} - k_2 (x_{2,\mathrm{ref},n} - x_{2e,n}) \end{bmatrix} \tag{6.31}$$

and equations (6.23) and (6.25) in (6.31)

$$\begin{bmatrix} e_{1,n+1} \\ e_{2,n+1} \end{bmatrix} = \begin{bmatrix} k_1 & T \\ 0 & k_2 \end{bmatrix} \begin{bmatrix} e_{1,n} \\ e_{2,n} \end{bmatrix} \tag{6.32}$$

$$\text{As } 0 < k_1, k_2 < 1 \quad \Rightarrow \quad (e_{1,n}, e_{2,n}) \quad \rightarrow (0,0), \quad n \quad \rightarrow \infty \tag{6.33}$$

6.6 Design of the Linear Algebra-Based Controller for a Linearized Model of the Continuous Stirred Tank Reactor

In Sect. 6.3, the LAB CD approach has been applied for a nonlinear model of a CSTR. As the reference does not have wide oscillations, a controller for an experimental linearized model is sought. Based on the model of the CSTR, (6.1)–(6.3) with the parameters listed in Table 6.1, a sudden increment in the refrigerating flow has been applied. At the initial stabilized operation, with a flow $q = 0.0033$ and a reactor concentration $C_a = 1.161$, a 20% increment in the flow is applied. The reaction curve is plotted in Fig. 6.6.

From this reaction curve, the FOPDT parameters are estimated (Sardella et al., 2019) as

$$G_1(s) = \frac{\Delta y(s)}{\Delta u(s)} = \frac{K}{\tau s + 1} \quad e^{-t_0 s}; \quad t_0 = 384.5; \quad \tau = 907.5; \quad K = 15.36$$

If the linearized LAB controller obtained following the procedure outlined in the previous section is applied, the obtained concentration response is depicted in Fig. 6.7.

In order to compare the results with those obtained with the controller developed based on the nonlinear model (6.11), the same oscillatory reference is required. The response of the CSTR with the controller designed based on the linear experimental model (6.26) is plotted in Fig. 6.8.

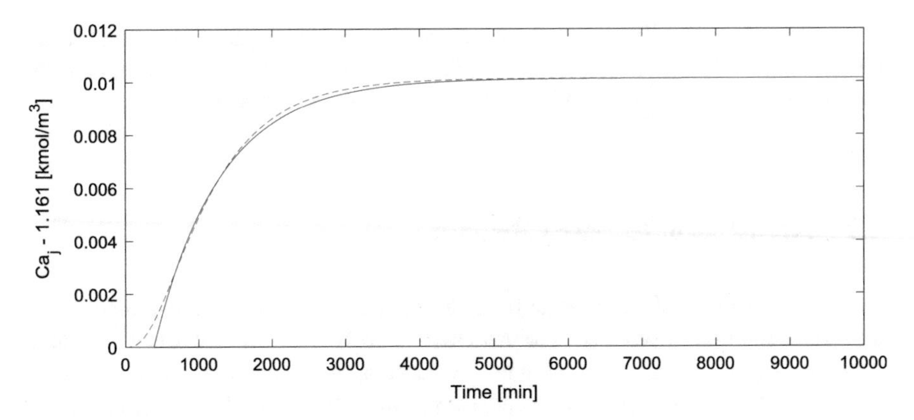

Fig. 6.6 Reactor reaction curve (dashed line) and FOPDT approximation

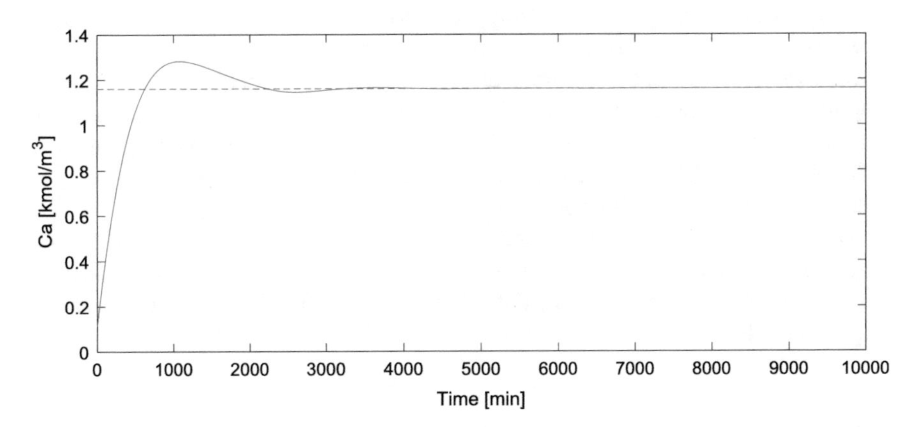

Fig. 6.7 Step response of the reactor (FOPDT controller)

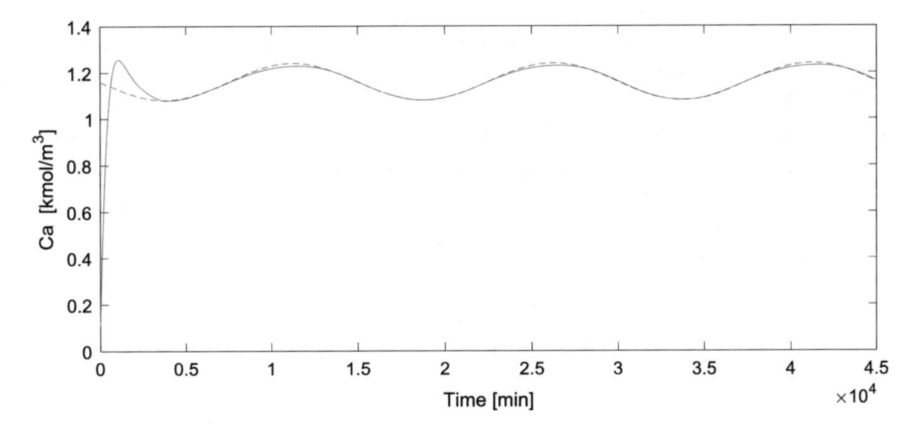

Fig. 6.8 Reactor response controlled by the linear model–based controller

It is worth noting that, other than the transient behavior, there is a small delay (as assumed in the approximated model) in the response. This delay does not appear in the case of the nonlinear LAB controller, as shown in Fig. 6.2.

6.7 Application to Batch Reactors

Batch processing is widely used in the industry, particularly in the manufacturing of goods and commodity products. The process state evolves from an initial to a final time, differing significantly from continuous operation, in which the process state is maintained at a desired operating point. They permit more flexibility, allowing adjustment of the operating conditions and the final time, but they require special attention regarding the coordination of different stage times and the determination of optimal temperature and feeding profiles. The application of the results obtained for continuous processes control cannot be transferred to batch processes due to the significant differences between them. Due to the finite characteristic of batch operation, stability loses importance, and reproducibility becomes more relevant. In this case, the problem of stability is, in fact, related to the sensitivity with respect to perturbations (Bonvin & Francois, 2017). Therefore, the main objective of the control systems must be to give fast set-point response without overshoot or oscillations.

In order to evaluate the proposed LAB CD methodology, a transesterification batch reactor from biodiesel production process at laboratory scale is considered. The results obtained by using a FOPDT model, as well as controlling the reactor experimentally, are reported.

6.7.1 *Experimental Batch Reactor*

Variable temperature profiles have proven to be the best operating conditions for many processes. A laboratory batch reactor was chosen to test LAB controller for variable temperature profiles. Experimental assays were performed in a jacketed glass reactor with an overhead stirrer (Fig. 6.9).

The response of the system was modeled by an FOPDT model, whose parameters were obtained from the reaction curve of the system to a 17% step change in the controller output (Fig. 6.10).

The resulting FOPDT model was:

$$G_{\exp}(s) = \frac{2.29}{75.75s + 1} \mathrm{e}^{-7.75s} \tag{6.34}$$

Fig. 6.9 Experimental reactor

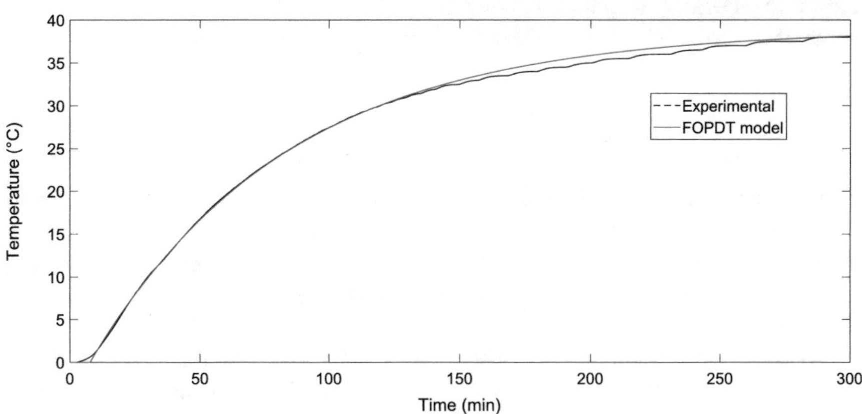

Fig. 6.10 Reaction curve of the experimental batch reactor

6.7.2 *Linear Algebra-Based Control Design*

The system was tested for constant and variable reference profiles using $k_1 = 0.95$, $k_2 = 0.94$. The tuning procedure used to find the controller parameters was based on a Monte Carlo algorithm, programed to minimize a proposed error index (Pantano et al., 2017; Tempo & Ishii, 2007; Rómoli, Serrano, Ortiz, Vega, & Scaglia, 2015; Fernández et al., 2019). This method allows estimating an expectation value and it provides effective tools for the analysis of probabilistically robust control schemes.

In this case it was applied to select the optimal controller parameters. In order to assure a bounded probability of an incorrect result, the minimum necessary number of trials (*N*) was determined, as described in Section 3.3. The method was applied as follows:

1. A confidence level (δ) and an accuracy degree (ε) are selected. Then the number of trials (N) is computed according to:

$$N \geq \left[\frac{\log \frac{1}{\delta}}{\log \frac{1}{1-\varepsilon}} \right] \quad (6.35)$$

   Typical values are $\delta = 0.01$ and $\epsilon = 0.005$. Therefore, $N \geq 920$.
2. Random values are selected for the controller parameters and a simulation of the controlled system is carried out.
3. An index used to evaluate the controller performance is calculated.
4. This procedure is repeated for *N* trials.
5. The optimal controller parameter set is the one for which the index is the best. In the cases in which the index reflects the tracking error, it should be the minimum.

The criterion to evaluate the performance was chosen from indexes that consider the entire close loop response. Among these, the integral of time weighted absolute value of the error (ITAE) is the best to eliminate errors that persist in time.

$$\mathrm{ITAE} = \int t.|e_1(t)|.dt \quad (6.36)$$

Experimental results showed that the LAB controller has a good performance, better than PID, when applied to a real process. Figure 6.11 shows the system following the reference signal even when a disturbance, caused by the addition of cold oil, appeared at minute 230. Moreover, the LAB controller corrected deviations faster than PID.

To assess the controller performance against parametric uncertainty, one test was carried out in a reactor with different model. The FOPDT assumed for this reactor was:

$$G_{\exp\_u}(s) = \frac{1.11}{55.6s + 1} \mathrm{e}^{-8.9s} \quad (6.37)$$

The new system has significant variations in model parameters: 51.5% in the steady-state process gain (K), 26.6% in the time constant (τ), and 14.8% in the dead time (t_0).

The process was tested using the LAB controller designed and tuned for the nominal system (6.34).

The system replies under variable profiles are shown in Fig. 6.12. Under nominal conditions, the LAB controller has an excellent performance. However, under

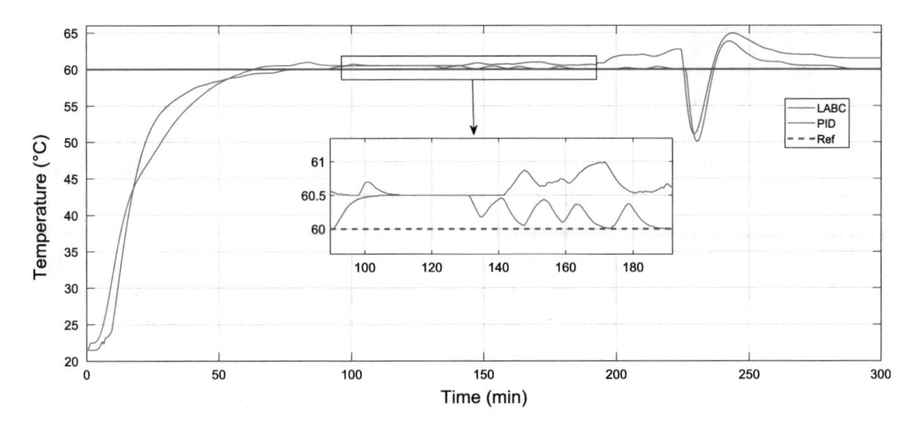

Fig. 6.11 Experimental results for LAB controller and PID. System responses against constant disturbances

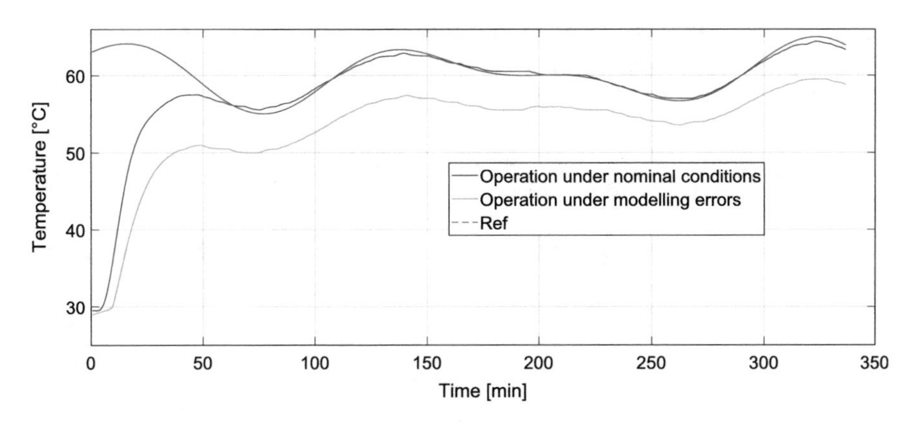

Fig. 6.12 Experimental results for LABC applied to variable profiles. Operation under nominal conditions and under modelling errors

modeling errors, the system is unable to correct some deviations, and a difference between the reference and the measured variable remains in time. The solution to this problem will be analyzed in Chap. 7.

6.8 Discussion about the Use of Linear Models

In this chapter, the use of linearized models to design the control has been introduced as an alternative to the use of nonlinear models, based on first principles. It is clear that there are some advantages and inconveniences. Nonlinear models lead to better controllers, although they require more complex computations. They also require the access to the full state of the plant or the implementation of state observers, involving additional complexity in the control structure.

Linearized models are very common in the industry, mainly when used to derive the control system. The main advantages are the simplicity of the model, also leading to simpler controllers, and the arbitrary definition of the state variables, allowing for an output derivative control implementation. The main drawback is the approximation inherent to the linearized model.

The simplified model can also be used if the initial model is linear but with high dimension. In some cases, a sort of *gain-scheduling* can be used to simplify the control design and implementation when dealing with nonlinear systems.

6.8.1 Linear High-Order System

The use of the LABC was also tested for a high-order system, being approximated by a FOPDT model. Let us consider the process defined by the following transfer function:

$$G_4(s) = \frac{64.103}{s^4 + 15.024s^3 + 70.112s^2 + 120.192s + 64.103} \mathrm{e}^{-0.5s} \tag{6.38}$$

From the reaction curve procedure (Liu et al., 2013), an FOPDT model was obtained:

$$G_5(s) = \frac{1}{1.3s + 1} \mathrm{e}^{-1.15s} \tag{6.39}$$

That means that the original model is a fourth-order system plus dead time (6.38), with a controllability relationship close to one ($\frac{t_0}{\tau} = 0.88$). The response curve of the system and the FOPDT model are shown in Fig. 6.13.

The system was tested for constant and variable reference profiles, using $k_1 = 0.9$, $k_2 = 0.7$.

The performance when using the LAB controller for a constant set point is shown in Fig. 6.14. Results are compared with the ones obtained using a PID controller. The best parameter values for the PID controller were found applying Monte Carlo algorithm (Pantano et al., 2017; Tempo & Ishii, 2007). Starting values were taken from Dahlin formula (Smith & Corripio, 1997) with a $\pm 20\%$ variation. The lowest ITAE was used for the selection.

The LAB controller leads the system to the reference faster than PID and without overshoot, mainly due to the feedforward component in the control law.

The LABC was also tested for a variable time reference and compared with that of the PID controlled plant (Fig. 6.15). The better performance when using the LABC is remarkable.

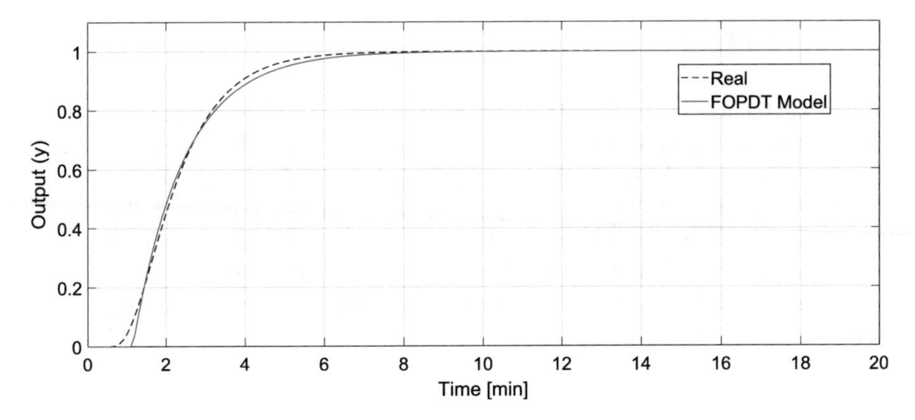

Fig. 6.13 Reaction curve

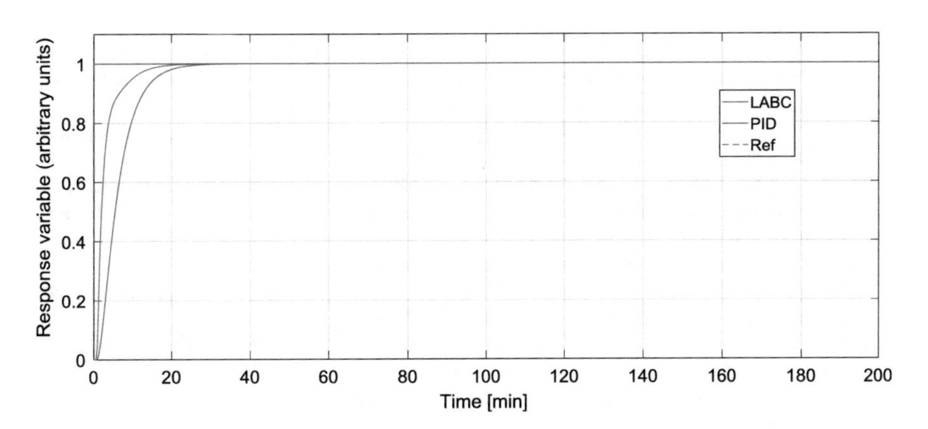

Fig. 6.14 Simulation results for a fourth-order system plus dead time for constant set point

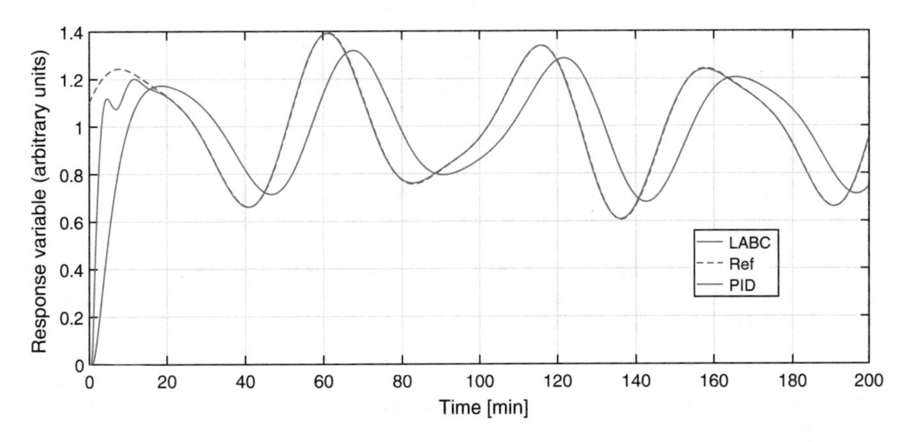

Fig. 6.15 Simulation results for a fourth-order system plus dead time for a variable profile

6.8.2 Piece-Wise Linearized Model

The linearized model (6.12) is only valid to represent the process behavior in a small region around the equilibrium point. If the reference to be tracked moves far from a given point, the model parameters can be adapted, in such a way that a battery of linearized models can be used along the trajectory. If this is the case, the control action computed by (6.26) should be adapted, according to the control action parameters given in (6.15) as a function of the model parameters.

Further research in this control strategy is being carried out, and preliminary results are very promising.

References

Bonvin, D., & Francois, G. (2017). Control and optimization of batch chemical processes. *Coulson and Richardson's Chemical Engineering*. Volume 3 (Chemical & Biochemical Reactors, and Process Control), 4th Edition, by J.F. Richardson and D.G. Peacock (Eds).

Camacho, O. A., & Smith, C. (2000). Sliding mode control: An approach to regulate nonlinear chemical processes. *ISA Transactions, 39*, 205–218.

Fern´ndez, M. C., Pantano, M. N., Rossomando, F., Ortiz, O. A., Scaglia, G., & Scaglia, J. E. (2019). State estimation and trajectory tracking control for a nonlinear and multivariable bioethanol production system. *Brazilian Journal of Chemical Engineering, 36*(1), 421–437. https://doi.org/10.1590/0104-6632.20190361s20170379.

Coughanowr, C., Ansara, I., Luoma, R., Hamalainen, M., & Lukas, H. L. (1991). Assessment of the Cu-Mg system. *Zeitschrift für Metallkunde, 82*(7), 574–581.

Liu, T., Wang, Q. G., & Huang, H. P. (2013). A tutorial review on process identification from step or relay feedback test. *Journal of Process Control, 23*(10), 1597–1623.

Luyben, W. L. (1990). *Process modeling, simulation and control for chem. engineers*. Singapore: McGraw-Hill.

Pantano, M. N., Serrano, M. E., Fernandez, M. C., Rossomando, F. G., Ortiz, O. A., & Scaglia, G. J. E. (2017). Multivariable control for tracking optimal profiles in a nonlinear fed-batch bioprocess integrated with state estimation. *Industrial and Engineering Chemistry Research, 56*, 6043–6056.

Ray, W.H. (1981). New approaches to the dynamics of nonlinear systems with implications for process and control system design. United States: N. p., 1981. Web.

Rómoli, S., Serrano, M. E., Ortiz, O. A., Vega, J. R., & Scaglia, G. J. E. (2015). Tracking control of concentration profiles in a fed-batch bioreactor using a linear algebra methodology. *ISA Transactions, 57*, 162–171. https://doi.org/10.1016/j.isatra.2015.01.002.

Sardella, M. F., Serrano, M. E., Camacho, O., & Scaglia, G. (2019). Design and application of a linear algebra based controller from a reduced-order model for regulation and tracking of chemical processes under uncertainties. *Industrial & Engineering Chemistry Research Publisher: American Chemical Society, 1*, 2019. https://doi.org/10.1021/acs.iecr.9b01257.

Seborg, D. E., Edgar, T. F., Mellichamp, D. A., & Doyle, F. J., III. (2011). *Process dynamics and control* (3rd ed.). Hoboken, NJ: Wiley.

Smith, C. A., & Corripio, A. B. (1997). *Principles and practice of automatic process control*. Hoboken, NJ: Wiley.

Stephanopoulos, G., & Vallino, J. J. (1991). Network rigidity and metabolic engineering in metabolite overproduction. *Science, 252*(5013), 1675–1681.

Tempo, R., & Ishii, H. (2007). Monte Carlo and Las Vegas randomized algorithms for systems and control. *European Journal of Control, 13*, 189–203.

Chapter 7
Uncertainty Treatment

In this chapter, the problem of uncertainty in the model will be considered. Some ideas were already presented in the previous chapters but here both the marine vessel treated in Chap. 5 and the batch reactor considered in the previous chapter will be analyzed in detail. These two processes may be considered as representative of a broad class of processes where the Linear Algebra-Based Control Design methodology can be applied.

In logistic and transport applications, the load change when a vehicle is tracking a desired trajectory makes maneuvering tasks a growing problem. The control of the course of the vehicle directly influences its maneuverability, the safety of navigation, and the time of arrival at the destination. To deal with this problem in the literature, classical control schemes have been improved by incorporating robust and adaptable control techniques (Scaglia, Mut, Jordan, Calvo, & Quintero, 2009). An improvement developed for linear algebra design method is presented in this chapter, in order to reduce the effect in the tracking error when environmental uncertainties appear.

In the reference book *Robust and Optimal Control* (Zhou, Doyle, & Glover, 1996, Chap. 9), a clear statement when dealing with the modeling and control of a real system can be read:

"Most control designs are based on the use of a design model. The relationship between models and the reality they represent is subtle and complex. A mathematical model provides a map from inputs to responses. The quality of a model depends on how closely its responses match those of the true plant. Since no single fixed model can respond exactly like the true plant, we need, at the very least, a set of maps. However, the modeling problem is much deeper; the universe of mathematical models from which a model set is chosen is distinct from the universe of physical systems. Therefore, a model set which includes the true physical plant can never be constructed. It is necessary for the engineer to make a leap of faith regarding the applicability of a particular design based on a mathematical model. To be practical, a design technique must help make this leap small by accounting for the inevitable inadequacy of models. A good model should be simple enough to facilitate the

G. Scaglia et al., *Linear Algebra Based Controllers*,
https://doi.org/10.1007/978-3-030-42818-1_7

design, yet complex enough to give the engineer confidence that designs based on the model will work on the true plant."

Based on these ideas, the systematic approach applied to deal with uncertainty in this chapter relies on considering an additive uncertainty and design the control to reduce its effect, keeping the main properties of the LAB controller: simplicity of the control law and easy understanding of the control solution. First, the uncertainty treatment is presented, and later on it is applied to these two typical processes, a marine vessel and a reactor.

7.1 Model Uncertainty

Assume a discrete time model of the process to be controlled, like (Scaglia et al., 2009)

$$x_{n+1} = \overline{h}(x_n, u_n) + E_n \tag{7.1}$$

where E_n is an additive uncertainty, with the same dimension as the state vector, that is, $E_n \in R^r$. Notice that the additive uncertainty can be used to model perturbed systems as well as a wide class of model mismatches. If the actual plant is precisely represented by $h(x_n, u_n)$, the additive uncertainty can be expressed by

$$E_n = h(x_n, u_n) - \overline{h}(x_n, u_n) \tag{7.2}$$

Note that if, as it will be assumed, x and u are bounded and h is Lipschitz, then E_n can be modeled as a bounded uncertainty (Mayne, Rawlings, Rao, & Scokaert, 2000; Michalska & Mayne, 1993).

In order to simplify the disturbance treatment, each element of the disturbance vector is going to be considered as a polynomial of similar order, such that the disturbance can be expressed as

$$E_n = \left[\sum\nolimits_{j=0}^{d} c_{ij} n^j\right] \tag{7.3}$$

According to the order of the polynomial, the disturbance can be counteracted with some additional actions. Let us consider the elements of the disturbance vector, one by one.

For instance, if the disturbance is constant, $E_n = \text{const.} = c_0$, then $d = 0$. This implies that the disturbance variation is null.

$$\delta E_n = E_n - E_{n-1} = 0 \tag{7.4}$$

In a similar way, for a first-order disturbance, $E_n = c_0 + c_1 n$, the second difference vanishes, that is

$$\delta^2 E_n = \delta E_n - \delta E_{n-1} = c_1 - c_1 = 0 \tag{7.5}$$

and as a rule, the q-th order difference is defined as $\delta^q E_n = \delta(\delta^{q-1} E_n)$. It is clear that for a polynomial disturbance as defined in (7.3), the differences are such that $\delta^d E_n = \text{const}; \quad \delta^{d+1} E_n = 0$.

Remark 7.1 The q-th difference of a $q-1$ order polynomial is zero.

7.1.1 Integral Action

In order to reduce the effect of E_n, some integrators of the tracking errors in the system state variables will be introduced, depending on the assumed time variation of E_n. In this way, if there is a constant uncertainty, (7.4), in the i-state variable, an integrator placed at the error $e_{i,n} = x_{i,\text{ref},n} - x_{i,n}$ will force the error to converge to zero.

Assuming a constant disturbance in all the state variables, the accumulative error in the state vector e_n will be described by

$$\mathbf{U}_{1,n+1} = \mathbf{U}_{1,n} + Te_n \tag{7.6}$$

Thus, the control action will be computed, assuming a new term based on the integral term. That is, in the definition of the controller, it is proposed to redefine eq. (2.29) incorporating the integral term,

$$x_{n+1} = x_{\text{ref},n+1} - k\underbrace{(x_{\text{ref},n} - x_n)}_{e_n} + K_1 \mathbf{U}_{1,n+1} \tag{7.7}$$

where k, K_1 are the proportional and integral control parameters, respectively.

Following a similar reasoning to the uncertainty treatment in Chap. 2 (2.36), now in DT, the error dynamics will be expressed as

$$e_{n+1} = ke_n - K_1 \mathbf{U}_{1,n+1} - E_n \tag{7.8}$$

That is, taking into account (7.4) and (7.6)

$$e_{n+2} + (-k + K_1 T - 1)e_{n+1} + ke_n = -\underbrace{(E_n - E_{n+1})}_{\delta E_{n+1} = 0} \tag{7.9}$$

which determines the controller parameters to get a stable behavior.

7.1.2 Double Integral Actions

Let us now consider that the uncertainty can be modeled by a function where the second-order difference is zero, such that $\delta^2 E_n = \delta(\delta E_n) = \delta(E_{n+1} - E_n) = E_{n+2} - 2E_{n+1} + E_n = 0$. Then, a double integrator should be introduced in a similar way to (7.7), defining the integrating variables$\mathbf{U}_1$, $\mathbf{U}_2$.

$$\mathbf{U}_{2,n+1} = \mathbf{U}_{2,n} + T\mathbf{U}_{1,n+1} \tag{7.10}$$

In this case, the control action will be computed, assuming an additional term in (7.7), such as

$$x_{n+1} = x_{\text{ref},n+1} - k\underbrace{(x_{\text{ref},n} - x_n)}_{e_n} + K_1\mathbf{U}_{1,n+1} + K_2\mathbf{U}_{2,n+1} \tag{7.11}$$

where k, K_1, K_2 are, respectively, the proportional, integral, and double integral control parameters. Operating as before, and taking into account that $\delta^2 E_n = 0$, the error dynamics can be expressed by

$$\begin{aligned} &e_{n+3} + (-k + T(K_1 + TK_2) - 2)e_{n+2} + (2k - K_1T + 1)e_{n+1} - ke_n \\ &= -\underbrace{\delta^2 E_{n+1}}_{=0} \end{aligned} \tag{7.12}$$

Now, as can be seen in (7.12), under the assumption of constant or linear varying uncertainty, $\delta^2 E_n = 0$, the uncertainty has no influence on the error dynamics. The controller parameters k, K_1, K_2 are chosen, in order to ensure the stability of the linear system represented in the left part of (7.12), as shown in the previous case.

7.1.3 Multiple Integral Actions

Following a similar reasoning, if the uncertainties can be approximated with a $q-1$ order polynomial, the influence of E_n on e_n will be eliminated by introducing q integrators. Consider that the uncertainty can be modeled by a function where the *q-order* difference is zero, such that $\delta^q \mathbf{E}_n = \delta(\delta^{q-1}\mathbf{E}_n) = 0$. Then, a q-integrator should be introduced in a similar way to (7.11), defining the integrating variables $\mathbf{U}_1$, $\mathbf{U}_2, \ldots, \mathbf{U}_q$.

$$\mathbf{U}_{q,n+1} = \mathbf{U}_{q,n} + \int_{nT}^{(n+1)T} \mathbf{U}_{q-1}(t)dt \cong \mathbf{U}_{q,n} + T\mathbf{U}_{q-1,n+1} \tag{7.13}$$

In this case, the control action will be computed, assuming q additional terms, such as:

$$x_{n+1} = x_{\text{ref},n+1} - k\underbrace{(x_{\text{ref},n} - x_n)}_{e_n} + \sum_{i=1}^{q} K_i \mathbf{U}_{i,n+1} \tag{7.14}$$

where k, K_1, K_2, ..., K_q are the proportional and integral control actions, respectively. Operating as before, and taking into account that $\delta^q E_n = 0$, the error dynamics can be expressed by

$$\mathbf{e}_{n+q+1} + \sum_{i=1}^{q-1} f_i(k, K_1, \ldots, K_q, T)e_{n+1} - ke_n = |B| - \underbrace{\delta^q E_n}_{=0} \tag{7.15}$$

where $|B|$ is a bounded nonlinearity. In (7.15), under the assumption of $\delta^q \mathbf{E}_n = \mathbf{0}$, the uncertainty has no influence on the error dynamics. The controller parameters, k, K_1, K_2, . . ., K_p are chosen, in order to ensure the stability of the linear system represented in the left part of (7.15), as shown in the earlier cases.

7.2 Controller Design for a Marine Vessel

Now, dealing with two of the previously studied processes, an additive uncertainty is incorporated and the LAB control design is illustrated. First, the model of the marine vessel considered in Chap. 5 (5.4) is revisited. The following uncertain model (Serrano, Godoy, Gandolfo, Mut, & Scaglia 2018) is assumed

$$\begin{bmatrix} x_{n+1} \\ y_{n+1} \\ \psi_{n+1} \\ v_{n+1} \\ u_{n+1} \\ r_{n+1} \end{bmatrix} = \begin{bmatrix} x_n \\ y_n \\ \psi_n \\ v_n \\ u_n \\ r_n \end{bmatrix} + T \begin{bmatrix} u_n \cos\psi_n - v_n \sin\psi_n \\ u_n \sin\psi_n + v_n \cos\psi_n \\ r_n \\ -\frac{m_{11}}{m_{22}} u_n\, r_n - \frac{d_{22}}{m_{22}} v_n \\ \frac{m_{22}}{m_{11}} v_n\, r_n - \frac{d_{11}}{m_{11}} u_n \\ \frac{m_{11} - m_{22}}{m_{33}} v_n\, u_n - \frac{d_{33}}{m_{33}} r_n \end{bmatrix} + \begin{bmatrix} 0 & 0 \\ 0 & 0 \\ 0 & 0 \\ 0 & 0 \\ \frac{1}{m_{11}} & 0 \\ 0 & \frac{1}{m_{33}} \end{bmatrix} \begin{bmatrix} \tau_{u,n} \\ \tau_{r,n} \end{bmatrix} + E_n;$$

$$E_n = \begin{bmatrix} E_{x,n} \\ E_{y,n} \\ E_{\psi,n} \\ E_{v,n} \\ E_{u,n} \\ E_{r,n} \end{bmatrix} \tag{7.16}$$

where E_n is the additive uncertainty.

Now, the LAB CD methodology is applied without considering the disturbance, and the effect of the uncertainty is analyzed. Replacing $\tau_{u,\,n}$ and $\tau_{r,\,n}$ from (5.15) in (7.16), after some simple operations, it yields

$$e_{n+1} = \begin{bmatrix} e_{x,n+1} \\ e_{y,n+1} \\ e_{\psi,n+1} \\ e_{u,n+1} \\ e_{r,n+1} \end{bmatrix} = k\ e_n + T \begin{bmatrix} h_{1,n} \\ h_{2,n} \\ h_{3,n} \\ 0 \\ 0 \end{bmatrix} - \begin{bmatrix} E_{x,n} \\ E_{y,n} \\ E_{\psi,n} \\ E_{u,n} \\ E_{r,n} \end{bmatrix} \tag{7.17}$$

Looking at (7.17), a direct effect of the additive uncertainty on the tracking error can be seen. To reduce its effect, an integral action is devised. That means that the control structure will be similar to (7.7)

$$\begin{bmatrix} x \\ y \\ \psi \\ u \\ r \end{bmatrix}_{n+1} = \begin{bmatrix} x \\ y \\ \psi \\ u \\ r \end{bmatrix}_{\text{ref},n+1} - \begin{bmatrix} k_x e_{x_n} \\ k_y e_{y_n} \\ k_\psi e_{\psi_n} \\ k_u e_{u_n} \\ k_r e_{r_n} \end{bmatrix} + \begin{bmatrix} K_{1x} U_{1,x,n+1} \\ K_{1y} U_{1,y,n+1} \\ K_{1\psi} U_{1,\psi,n+1} \\ K_{1u} U_{1,u,n+1} \\ K_{1r} U_{1,r,n+1} \end{bmatrix} \tag{7.18}$$

where the integral terms are computed as in (7.6) and the controller coefficients are chosen to ensure the Hurwitz character of the polynomial in (7.7).

Once this integral term has been added, the LAB CD methodology is applied, following the same procedure applied to obtain (5.11), (5.12), and (5.13), in order to compute the new heading angle, forward velocity, and angular velocity. Thence, the heading angle will be computed as

$$\begin{aligned}\tan\psi_{\mathrm{ref},n} &= \frac{\sin\psi_{\mathrm{ref},n}}{\cos\psi_{\mathrm{ref},n}}\\ &= \frac{y_{\mathrm{ref},n+1}-k_y\left(y_{\mathrm{ref},n}-y_n\right)-y_n+K_{1,y}U_{1,y,n+1}-Tv_n\ \cos\psi_n}{x_{\mathrm{ref},n+1}-k_x\left(x_{\mathrm{ref},n}-x_n\right)-x_n+K_{1,x}U_{1,x,n+1}+Tv_n\ \sin\psi_n}\end{aligned} \tag{7.19}$$

The reference forward velocity is

$$\begin{aligned}u_{\mathrm{ref},n} &= \left(\frac{\Delta_{y,\mathrm{ref}}+K_{1,y}U_{1,y,n+1}}{T}-v_n\cos\psi_n\right)\sin\psi_{\mathrm{ref},n}\\ &\quad+\left(\frac{\Delta_{x,\mathrm{ref}}+K_{1,x}U_{1,x,n+1}}{T}+v_n\sin\psi_n\right)\cos\psi_{\mathrm{ref},n}\end{aligned} \tag{7.20}$$

The angular velocity that makes the tracking errors to tend to zero must be

$$r_{\mathrm{ref},n} = \frac{\psi_{\mathrm{ref},n+1}-k_\psi\left(\psi_{\mathrm{ref},n}-\psi_n\right)-\psi_n+K_{1,\psi}U_{1,\psi,n+1}}{T} \tag{7.21}$$

Finally, the new values of the control actions are obtained using least squares.

$$\begin{bmatrix}\tau_{u,n}\\ \tau_{r,n}\end{bmatrix}=\begin{bmatrix} m_{11}\left(\dfrac{u_{\mathrm{ref},n+1}-k_u\left(u_{\mathrm{ref},n}-u_n\right)-u_n+K_{1,u}U_{1,u,n+1}}{T}-\dfrac{m_{22}}{m_{11}}v_n r_n+\dfrac{d_{11}}{m_{11}}u_n\right)\\ m_{33}\left(\dfrac{r_{\mathrm{ref},n+1}-k_r\left(r_{\mathrm{ref},n}-r_n\right)-r_n+K_{1,r}U_{1,r,n+1}}{T}-\dfrac{m_{11}-m_{22}}{m_{33}}v_n u_n+\dfrac{d_{33}}{m_{33}}r_n\right)\end{bmatrix} \tag{7.22}$$

Replacing the control actions ($\tau_{u,\ n}$; $\tau_{r,\ n}$) of (7.22) in (7.16), and after some simple operations, it yields

$$e_{n+2}+(-k+K_1T-1)e_{n+1}+ke_n = T\underbrace{\begin{bmatrix}h_{1,n+1}-h_{1,n}\\ h_{2,n+1}-h_{2,n}\\ h_{3,n+1}-h_{3,n}\\ 0\\ 0\end{bmatrix}}_{\text{Bounded nonlinearity}}-\underbrace{\left(E_{n+1}-E_n\right)}_{\delta E_n=0} \tag{7.23}$$

Therefore, k, K_1 are chosen in order to ensure the stability of the linear system represented in the left part of (7.23), that is, the zeros of this polynomial should be inside the unit circle. Then $\sqrt{e_{x,n}^2 + e_{y,n}^2} \to 0$, as $n \to \infty$. In this way, the tracking error tends to zero despite uncertainties, if they are constant.

A similar reasoning is developed if the disturbances are higher-order polynomials, introducing additional integrators. The obtained control laws will be denoted as C0 (5.15) if there are not integrators; C1 (7.22) if there is a line of integrators; and C2 if two integrators are added for each error.

7.2.1 Simulation Results

To assess the performance of the developed controller, disturbances are incorporated into the continuous model of the marine vessel, according to Fig. 7.1 and equation (7.24). The time variation of d_x and d_y is given in Fig. 7.2. As can be seen, first, a constant disturbance is introduced in both variables at time $t = 40$ s. Then, a linear function is added to the disturbance function at time $t = 85$ s, and finally, a quadratic signal is introduced at $t = 170$ s.

$$\begin{aligned}
\dot{x} &= u \ \cos\psi\text{-v} \ \sin\psi + d_x \\
\dot{y} &= u \ \sin\psi + \text{v} \ \cos\psi + d_y \\
\dot{\psi} &= \text{r} \\
\dot{u} &= f_{1,n} + \frac{b_{11}}{m_{11}}\tau_u \\
\dot{r} &= \frac{m_{22}}{m_{22}m_{33} - {m_{23}}^2}\left(f_{2,n} + b_{32}\tau_r\right)
\end{aligned} \tag{7.24}$$

The integrative controllers are implemented with zero (C0), one (C1), and two (C2) integrators, in order to analyze the performance according to the number of integrators. The parameter set for each controller is shown in Table 7.1.

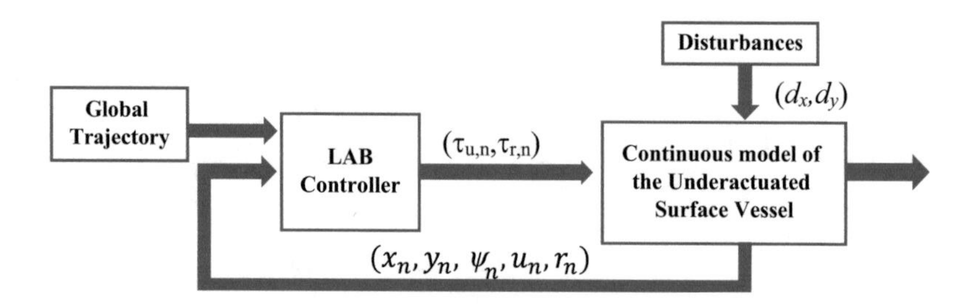

Fig. 7.1 Architecture of the trajectory tracking controller

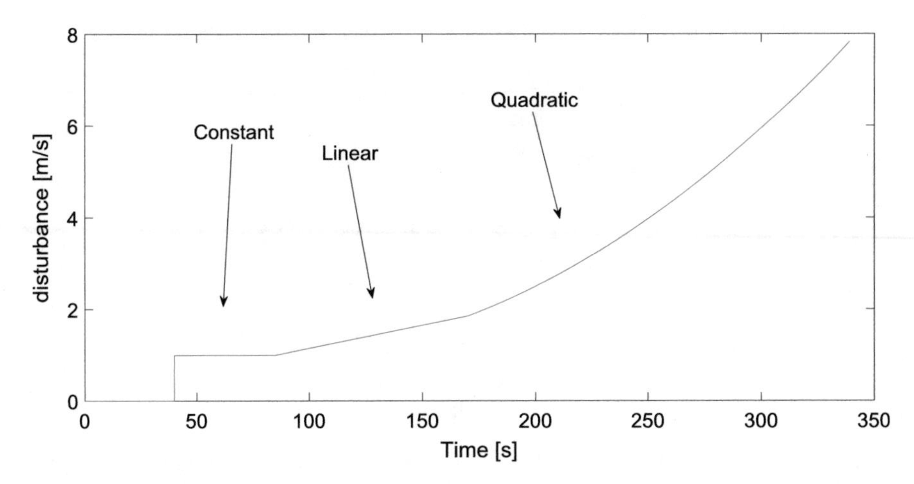

Fig. 7.2 Time variation of the disturbance introduced in the ship model

Table 7.1 Parameters set for each controller

Controller\results	Controllers parameters
C0	$k = 0.883$
C1	$k = 0.883$ $K_1 = 0.0269$
C2	$k = 0.883$ $K_1 = 0.0269$ $K_2 = 0.0011$

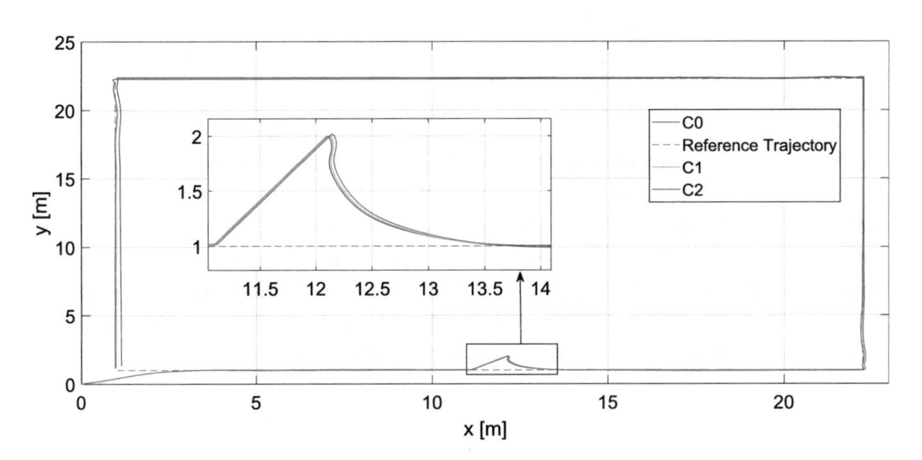

Fig. 7.3 Ship trajectory. Red dotted line: reference trajectory; blue line: C0 trajectory; green line: C1 trajectory; magenta line: C3 trajectory

The control performance when a square reference trajectory is given as the desired path is shown in Fig. 7.3. As it can be seen, all controllers reach and follow the desired trajectory when disturbances are introduced in the ship model. Time

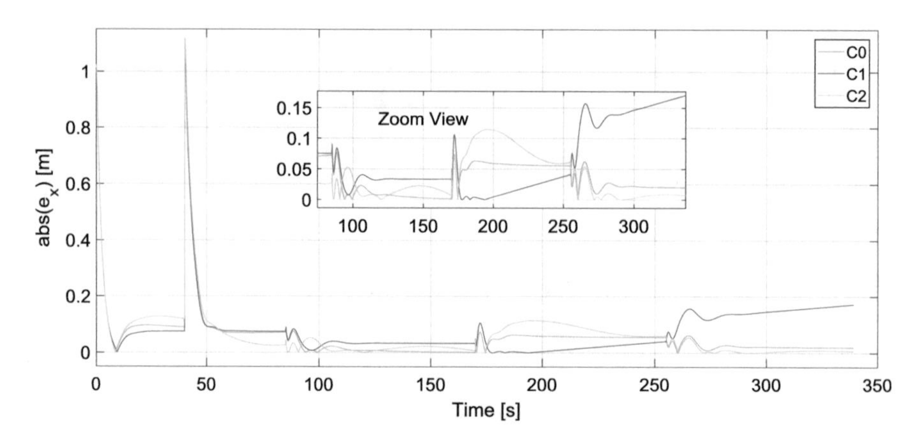

Fig. 7.4 Absolute error in x-variable. Blue line: C0 error; green line: C1 error; cyan line: C3 error

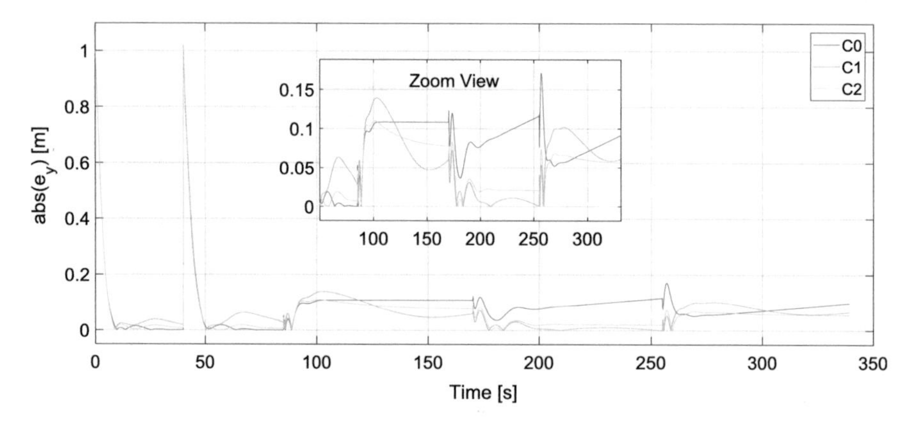

Fig. 7.5 Absolute error in y-variable. Blue line: C0 error; green line: C1 error; cyan line: C3 error

variation of the tracking errors is shown in Figs. 7.4 and 7.5. By inspection, it can be seen that the controller C2 provides the best performance when a polynomial disturbance is introduced. In addition, when a quadratic disturbance is introduced, the tracking error by using C0 linearly increases, while the error remains constant and close to zero if C1 and C2 are used, respectively.

7.3 Controller Design Under Uncertainty: Batch Reactor

As a second process to investigate the control design under uncertainty, a batch reactor is considered. As discussed in the previous chapter, it is rather frequent to approximate the behavior of industrial processes by FOPDT models, which is also an approximation of the real system. Thus, it is important to consider the presence of model uncertainty in the controller design (Mayne et al., 2000; Michalska & Mayne, 1993).

In this section, an analysis for the perturbed systems is performed, assuming the disturbances as well as model mismatches as transient ones. An unknown additive uncertainty is introduced into the model of the system, and (6.19) takes the form:

$$\begin{bmatrix} x_{1,n+1} \\ x_{2,n+1} \end{bmatrix} = \begin{bmatrix} x_{1,n} \\ x_{2,n} \end{bmatrix} + T\left\{ \begin{bmatrix} 0 & 1 \\ -K_B & -K_A \end{bmatrix} \begin{bmatrix} x_{1,n} \\ x_{2,n} \end{bmatrix} + \begin{bmatrix} 0 \\ KK_B \end{bmatrix} u_{1,n} \right\} + \begin{bmatrix} 0 \\ 1 \end{bmatrix} E_n \tag{7.25}$$

If the disturbances are not considered, the so-called proportional controller leads to the disturbed error equation (6.32), that is,

$$\begin{bmatrix} e_{1,n+1} \\ e_{2,n+1} \end{bmatrix} = \begin{bmatrix} k_1 & T \\ 0 & k_2 \end{bmatrix} \begin{bmatrix} e_{1,n} \\ e_{2,n} \end{bmatrix} + \begin{bmatrix} 0 \\ 1 \end{bmatrix} E_n \tag{7.26}$$

showing the effect of the disturbance in the tracking errors.

Note that the initial state of variable y_1 (called y_0 in Chap. 6), is no longer subtracted. It is assumed as an uncertainty to be compensated by the controller (Sardella, Serrano, Camacho, & Scaglia, 2019).

The mathematical formulation of the uncertainty that represents model mismatches can be done as follows:

Starting from the equation of the system (6.12):

$$\tau\dot{y}(t+t_0) + y(t+t_0) = K\,u(t) \tag{7.27}$$

In order to deal with the delay, a Taylor approximation (see Appendix A.4.1) is applied. Thence,

$$y(t+t_0) = y(t) + \dot{y}(t)t_0 + \underbrace{\ddot{y}(t+\lambda t_0)\frac{{t_0}^2}{2}}_{\text{Complementary term}=H\ (t)} \quad ; \quad 0<\lambda<1 \tag{7.28}$$

and

$$\dot{y}(t+t_0) = \dot{y}(t) + \ddot{y}(t)t_0 + \underbrace{\dddot{y}(t+\lambda t_0)\frac{{t_0}^2}{2}}_{\text{Complementary term}=\frac{dH(t)}{dt}} \quad ; \quad 0<\lambda<1 \tag{7.29}$$

replacing (7.28) and (7.29) into (7.27), the following expression is obtained:

$$\tau t_0 \ddot{y}(t) + (\tau + t_0)\dot{y}(t) + y(t) = Ku(t) - \left\{ H(t) + \tau \frac{dH(t)}{dt} \right\} \tag{7.30}$$

and taking into account (6.15), it yields

$$\ddot{y} + K_A \dot{y} + K_B y + K_B \left(H(t) + \tau \frac{H(t)}{dt} \right) = K K_B u \tag{7.31}$$

Thus, applying Euler DT approximation, (7.25) is obtained, and the uncertainty term takes the form:

$$E_n = -K_B T \left(H(nT) + \tau \left. \frac{dH(t)}{dt} \right|_{t=nT} \right) \tag{7.32}$$

By this term, all the model uncertainties are collected, including the time delay approximation, the uncertainty in the initial model parameters, as well as some additive extra disturbances.

Applying the LAB CD methodology to (7.25), including an integral action, the control law is:

$$u_n = \frac{1}{K K_B} \times \left(\frac{y_{2\mathrm{ref},n+1} - k_2 \left(y_{2\mathrm{ref},n} - y_{2,n} \right) - y_{2,n} + K_1 U_{n+1}}{T} + K_B y_{1,n} + K_A y_{2,n} \right) \tag{7.33}$$

where

$$U_{n+1} = U_n + \int_{nT}^{(n+1)T} e_2(t)\, dt \cong U_n + e_{2,n} \cdot T \tag{7.34}$$

The controlled plant model is obtained by replacing (7.33) in (7.25):

$$y_{2n+1} = y_{2,n} + \left[\left(y_{2\mathrm{ref},n+1} - k_2 e_{2,n} - y_{2,n} + K_1 U_{n+1} - K_A y_{2,n} T - K_B y_{1,n} T \right) + K_A y_{2,n} T + K_B y_{1,n} T \right] + E_n$$

$$y_{2n+1} = y_{2,\mathrm{ref},n+1} - k_2 e_{2,n} + K_1 U_{n+1} + E_n$$

$$\underbrace{y_{2\mathrm{ref},n+1} - y_{2,n+1}}_{e_{2,n+1}} = k_2 e_{2,n} - K_1 U_{n+1} - E_n$$

And after some simple operations, it yields

$$e_{2,n+2} - (k_2 - K_1T + 1)\,e_{2,n+1} + k_2e_{2,n} = \underbrace{E_{n+1} - E_n}_{\delta E_n} \tag{7.35}$$

To ensure the stability of the linear system, the zeros of the characteristic polynomial of equation (7.35) should comply:

$$0 \le |r_i| < 1, i = \{1, 2\}$$

Then $e_{2,\ n\ +\ 1} \to 0$ as $n \to \infty$ and $e_{1,\ n\ +\ 1} = k_1\,e_{1,\ n} + Te_{2,\ n} \to 0$ as $n \to \infty$.

This means that the error will tend to zero despite uncertainties, if they are constant.

Remark 7.1 Note that the integral term, added to correct uncertainties effects, is defined over e_2 instead of e_1, to avoid overshoot in the response variable. When e_1 is used, the expression of $y_{2,\ \text{ref},\ n}$ should be defined as:

$$y_{2\text{ref},n} = \frac{y_{1\text{ref},n+1} - k_1e_{1,n} - y_{1,n} + K_1\,U_{n+1}}{T}$$

and $U_{n\ +\ 1}$ takes the form: $U_{n+1} = U_n + \int_{nT}^{(n+1)T} e_1(t)\,dt \cong \ U_n + e_{1,n}\cdot T$.

And the expression for e_1, when there are no perturbations on the system, is:

$$e_{1n+1} - k_1e_{1,n} + T\,e_{2n} - K_1\,U_{n+1} = 0 \tag{7.36}$$

Working under nominal conditions, e_1 and e_2 should tend to zero. So, to satisfy (7.25), the integral term should tend to zero too. To accomplish this condition, e_1 will take positive and negative values, leading always to an overshoot on y_1.

The integral over e_2 produces the same effect on y_2, which is the derivative of y_1, but a change in the derivative sign does not imply an overshoot on y_1. This improvement allows the adjustment of the integral term of the controller, avoiding overshoot in the response.

Experimental results demonstrate the superiority of this methodology, applied to the experimental batch reactor cited in Chap. 6. Figure 7.6 shows the system response for a variable time temperature profile, when the process is under nominal operation conditions and under uncertainties. These results are compared with the system controlled by the original LABC. It can be seen that the tracking error is very low when there are minor uncertainties, but it is unable to correct significant mismatches as the ones included in this work (Chap. 6). When the improved controller was used, the controller performance was as good as when no significant mismatches were present.

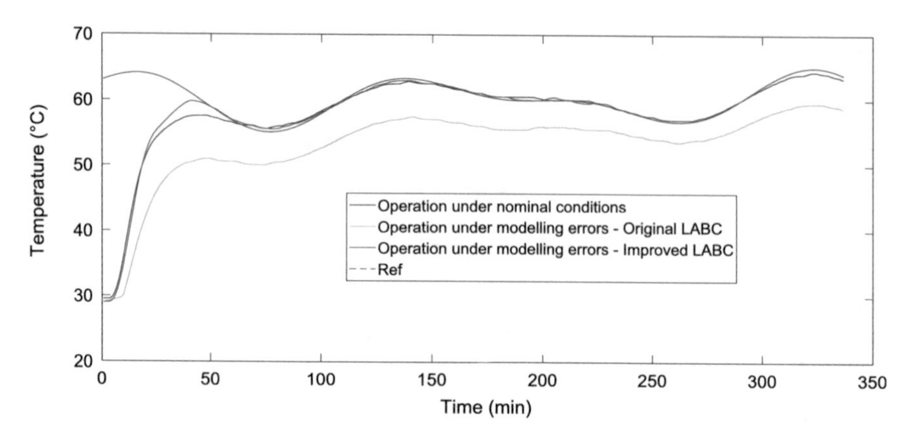

Fig. 7.6 Experimental results for LABC applied to the system under nominal and operation and under uncertainty

References

Mayne, D. Q., Rawlings, J. B., Rao, C. V., & Scokaert, P. O. M. (2000). Constrained model predictive control: stability and optimality. *Automatica, 36*, 789–814.

Michalska, H., & Mayne, D. Q. (1993). Robust receding horizon control of constrained nonlinear systems. *IEEE Transactions on Automatic Control, 38*, 1623–1633.

Sardella, M. F., Serrano, M. E., Camacho, O., & Scaglia, G. (2019). Design and application of a linear algebra based controller from a reduced-order model for regulation and tracking of chemical processes under uncertainties. *Industrial & Engineering Chemistry Research Publisher: American Chemical Society, 1*, 2019. https://doi.org/10.1021/acs.iecr.9b01257.

Scaglia, G. J. E., Mut, V. A., Jordan, M., Calvo, C., & Quintero, L. (2009). Mobile robot control based on robust control techniques. *Journal of Engineering Mathematics, 63*(1), 17–32. https://doi.org/10.1007/s10665-008-9252-0.

Serrano, M. E., Godoy, S. A., Gandolfo, D., Mut, V. A., & Scaglia, G. J. E. (2018). Nonlinear trajectory tracking control for marine vessels with additive uncertainties. *Information Technology and Control, 47*(1), 118–130. https://doi.org/10.5755/j01.itc.47.1.18021.

Zhou, K., Doyle, J. C., & Glover, K. (1996). *Robust and optimal control* (Vol. 40, p. 146). New Jersey: Prentice hall.

Chapter 8
Linear Algebra-Based Controller Implementation Issues

In the previous chapters, the Linear Algebra-Based Control Design methodology has been presented and applied to a number of processes, showing its applicability to a variety of models. But, obviously, this is not the panacea to design the control for any plant. There are some constraints but also some clear benefits.

In this chapter, the advantages in using this methodology to design the control will be first summarized and later on, some issues and critical drawbacks will be discussed, providing simple guidelines to implement the designed controller.

8.1 The Advantages

One of the motivations to develop the LAB CD methodology is related to its implementation facilities. In this sense, the main advantages of this procedure relay on:

- Easy derivation of the control law. The controller is obtained by computing a soft inverse model of the plant, with some constraints to ensure its realizability.
- The controller design options are simple: just the selection of the proportional controller parameters.
- The influence of the controller parameters is very apparent, leading to an easy-to-tune procedure.
- The control design procedure naturally leads to a feedback/feedforward control law.
- The methodology is equally applicable to linear and nonlinear plants with the model constraint that it should be affine in the control.
- The procedure is equally applicable for multivariable plants.
- Some model uncertainties, when considered as additive uncertainties, can be easily treated.

G. Scaglia et al., *Linear Algebra Based Controllers*,
https://doi.org/10.1007/978-3-030-42818-1_8

- There is a low control law computation load, just an algebraic computation, being easy to implement in low-cost digital systems.
- The designed controller ensures the stability and steady-state accuracy of the controlled plant.

Nevertheless, some drawbacks and warnings should be raised if this control design methodology is used. The following sections summarize the most relevant issues to be considered.

8.2 Sampling Period

No direct reference to the sampling period selection has been done previously. It is clear that, as in any discrete time controller, there is always a trade-off between the computational load and the control accuracy (both increase if the sampling period is reduced). Also, if the sampling period is enlarged, keeping the same controllers, the control actions will be larger and some saturation problems may appear. In fact, looking at the controller parameters, k_i, as defined in (2.27), they represent the percentage of error reduction in a sampling period. Thus, if the sampling period is shortened, these parameters should also be reduced to avoid strong control actions leading to actuators' saturation. One advantage of short sampling period is that the errors due to disturbances are detected earlier, and the disturbance effect can be reduced.

On the other hand, if the sampling period is larger, the control actions are smaller for the same controller parameters. Thus, they can be tuned easily, without much accuracy.

In order to illustrate this trade-off, let us consider the discrete time (DT) control design of the XY plotter introduced in Chap. 2 and review some performance measurements, as a function of the sampling period. The process model is repeated here (8.1), assuming the following parameters: $m = 1$; $r = 3$; $k = 2$; and $b = 1$.

$$m\ddot{y}(t) + r\dot{y}(t) + ky(t) = bf(t) \tag{8.1}$$

The system responses for the LAB controller defined for $k_1 = k_2 = 0.9$ and different sampling periods ($T = 0.1$, 0.05 s) are plotted in Fig. 8.1. The control actions are shown in Fig. 8.2.

In order to compare the effect of the sampling period, the following features are reported:

1. Square of the error norm:
   For $T = 0.1$ s $\|e\|^2 = 6.23766$; for $T = 0.05$ s $\|e\|^2 = 5.6205$
2. Control action. Zooming the initial time of the responses, the initial control action is greater for longer sampling period. That is, for shorter sampling periods, the control error is reduced, but the control action is increased.

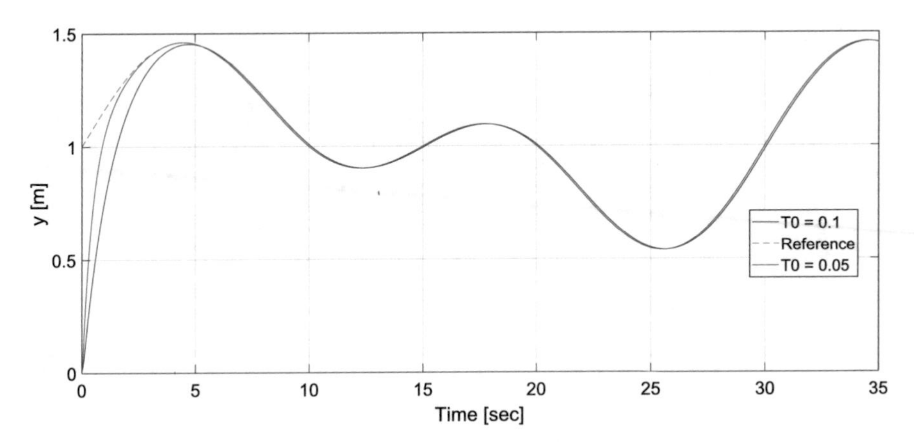

Fig. 8.1 System response for sampling periods $T = 0.1$ s and $T = 0.05$ s

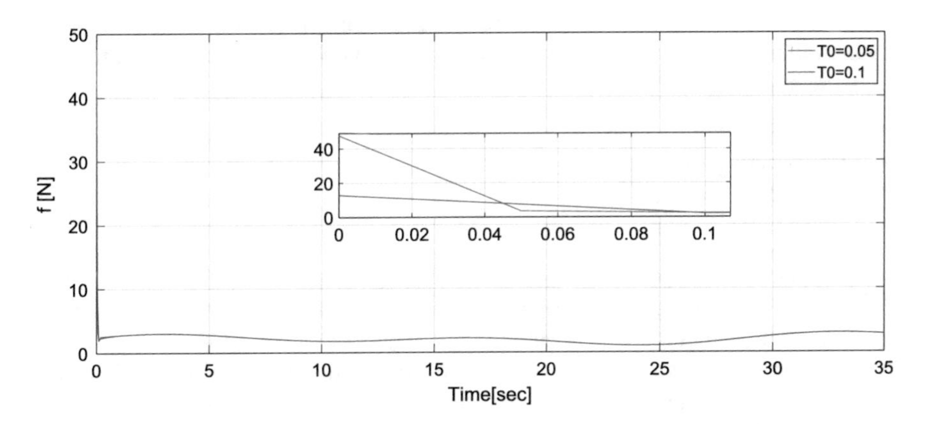

Fig. 8.2 Control action for $T = 0.1$ s and $T = 0.05$ s

3. It does not affect the complexity of the control, requiring the same time for the control computation.
4. To see the sampling period's influence on disturbance rejection, a change in the control action is assumed. The control has been designed for $b = 1.2$, but the parameter in the process is taken as $b = 1$. The responses for different sampling periods are plotted in Fig. 8.3. The tracking error is reduced as far as the sampling period is reduced.

8.3 Robustness

A simplified model is much better to design the control, but the degradation of performance is clear if there are errors in the model parameters.

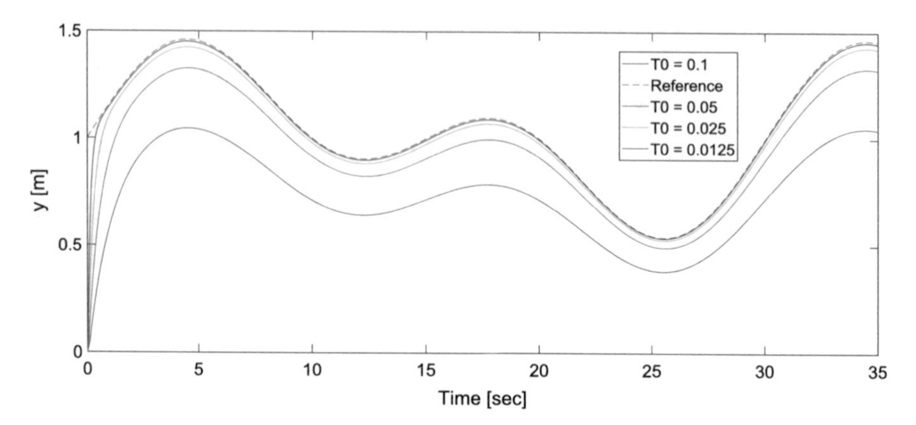

Fig. 8.3 System response under model mismatch and different sampling periods

In order to illustrate this degradation, let us consider different uncertainties in the industrial process model developed in Sect. 6.7 and review some performance measurements.

- Delay mismatch
- Steady-state gain mismatch
- Time delay mismatch
- External disturbance

As it can be seen in Sect. 6.7, when there are modeling errors (6.34, 6.37, and Fig. 6.12), the system performance degrades considerably. To solve this problem, it is proposed to add integrators about the error in the "sacrificed variable" (6.34). The response of the system with the new control action (6.33) can be seen in Fig. 6.6. Where it was possible to reduce the tracking error, without presenting overshooting, which would be the result if the integral would be calculated in the variable of interest (the tracking variable). It can be qualitatively seen that on adding integrators the system behaves as if there were no modeling errors. The clear disadvantage is that the order of the system has been increased and therefore the selection of the control parameters is more difficult.

8.4 Controller Parameter Tuning

In the previous control design applications, by using the LAB methodology, it has been illustrated that simple models and control structures lead to acceptable performance. Thus, the analysis will be limited to this control solution.

The typical approach to fix the controller parameters is to guarantee the controlled plant stability, and it has been proved that the option is clear and easy to implement. The problem is if some other features are required. Until now, no simple relationship between controller parameters and performance has been figured out. Thus, in the

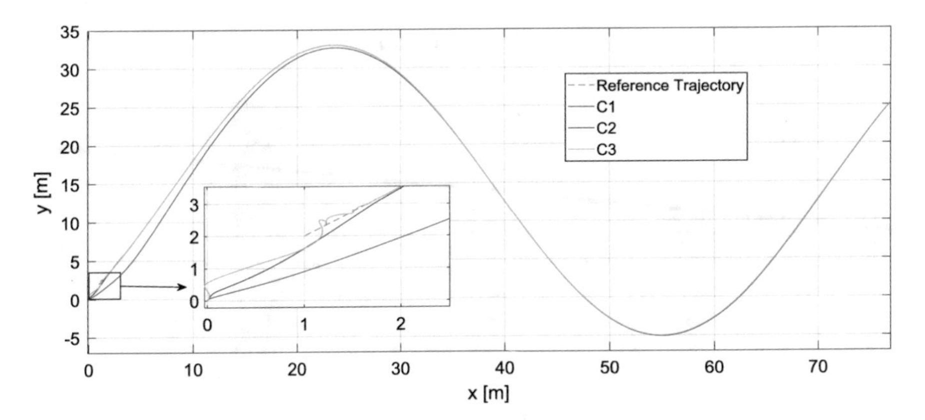

Fig. 8.4 Ship position for the three simulated cases

applications where some index should be optimized, a randomized optimization approach like the Monte Carlo optimization procedure has been applied.

Nevertheless, if the initial design does not fulfil the expected requirements, the end user should be capable to retune the controller parameters. In order to illustrate this facility, the basic control of a marine vessel developed in Chap. 5 will be designed when an error index is assumed as design criterion but, later on, on the implemented control, some parameters should be tuned to improve some visual performance like:

- Response time
- Oscillations
- Control effort

The simulations are carried out in MatLab SIMULINK. In order to compare the aforementioned features (Response time, Oscillations and Control effort), three simulations were performed, changing the parameters of the controller. The objective is to analyze the responses of the system when the controller's parameters vary; thus, three different sets of parameters were chosen. For the first case (C1), we chose $(k_x, k_y, k_\psi, k_u, k_r) = (0.999, 0.999, 0.98, 0.99, 0.99)$; for case 2 (C2) $(k_x, k_y, k_\psi, k_u, k_r) = (0.97, 0.97, 0.8, 0.92, 0.92)$; and, finally, for case 3 (C3), the following values were adopted $(k_x, k_y, k_\psi, k_u, k_r) = (0.51, 0.51, 0.35, 0.91, 0.91)$. The adopted sampling period was $T = 0.1$ s, and the reference trajectory was a sinusoidal path.

The position of the ship and the reference trajectory are shown in Fig. 8.4 for each simulation. As it can be seen, there are noticeable differences between the three responses. For C1, the response is smooth and the response time is the highest of the three cases. However, as can be seen in Fig. 8.5, the control effort is the smallest. When the values of the controller parameters decrease, the response time is lowering as can be seen for case 2. As can be seen in Fig. 8.4, for C2, the response does not show unwanted oscillations. However, the cost of this reduction is reflected with a slight increase in the control effort, Fig. 8.5. The C3 shows what happens when the controller parameters adopt very low values. In these cases, the system has a rapid

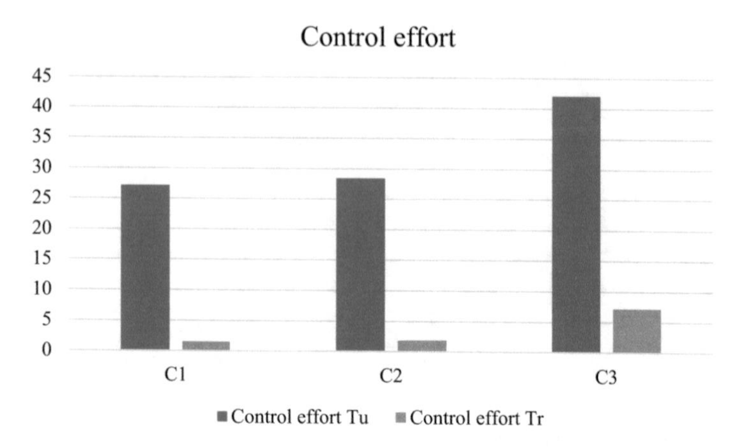

Fig. 8.5 Control effort for each simulation

response, but undesired oscillations may appear as can be seen in Fig. 8.4. In addition, another disadvantage if the controller parameters are too reduced is that the control effort significantly increases.

In order to facilitate the reader experimentation with this LAB control design methodology, a simulator has been arranged in which the values of the controller parameters can be changed. Thus, these conclusions as well as more advanced results can be directly obtained by the reader.

8.5 Trade-Off Simplicity Versus Performance

There are many control solutions for the tracking problem. Under the same conditions (model, requirements, reference), different solutions available in the literature are compared with the behavior of the plant controlled by a LAB controller. The basic mobile robot model will be used for this comparison.

The suggested approaches to compare with LAB, well referenced in the literature, are:

- Results based on the look-ahead methods (Das & Kar, 2006; Martins, Celeste, Carelli, Sarcinelli-Filho, & Bastos-Filho, 2008) use the feedback linearization technique. In this proposal, the intersection point of a straight line passing through the middle of the vehicle and an axis of the two wheels instead of the center of mass in the wheeled mobile robots is chosen in the configuration of the robot position. However, this approach has the following problem: as the distance between the center of mass and the intersection point becomes larger, the tracking performance will deteriorate. On the other hand, when it becomes smaller, the control input tends to become much larger as it involves the inverse of an almost singular matrix.

- In Resende, Carelli, & Sarcinelli-Filho (2013), a controller that combines the heuristic knowledge of the problem, the sector nonlinearity approach, and the inverse kinematic of the mobile platform was proposed. In Li, Chang, & Tong (2004), a fuzzy target tracking control unit (FTTCU), which comprises a behavior network for each action of the tracking control and a gate network for combining all the information of the infrared sensors, is used. The disadvantages of these methods rely on the amount of information that must be retrieved from the system, in order to construct the knowledge base for the control laws.
- In Sun & Cui (2004), a controller for trajectory tracking is designed, using the kinematic model of the mobile robot and a transformation matrix. Such matrix is singular if the linear velocity of the mobile robot is zero; therefore, the effectiveness of this controller is only assured if the velocity is different from zero.

The LAB control methodology allows the mobile robot to be controlled by solving the limitations of other controllers mentioned earlier. Other advantages that can be mentioned are as follows: it can be applied to the kinematic and dynamic model, in continuous time and in discrete time, and the controller tuning is easy. In addition, it offers good rejection of disturbances by incorporating integrators. However, one of the biggest advantages offered by the application of this methodology is that the lector does not need advanced knowledge of control theory to implement this technique, and that no change or transformation of variables is needed to solve the tracking problem.

8.6 Output Feedback

One of the LAB requirements is the access to the plant state. If there is a limited access to the state and the control is based on the reference and the plant output, an observer should be implemented, leading to a degrading in the performance. Of course, the solution largely depends on the plant model and the constraints in the state access (Fernández, Nadia Pantano, Rossomando, Ortiz, & Scaglia, 2019; Pantano et al., 2017; Rómoli et al., 2017; Rómoli, Amicarelli, Ortiz, Scaglia, & di Sciascio, 2016).

There is a lot of literature about the development of nonlinear observers (see, for instance, the book edited by Besançon, 2007). Of course, if the plant model has been linearized, a well-known linear observer such as the Luenberger observer or the Kalman filter can be easily implemented.

8.7 Process Limitations

The LAB CD methodology was inspired in designing the tracking control for a mobile robot. The model conditions and the control requirements were fulfilled, and the control law was easily derived. But, as already mentioned, this methodology is not the panacea to design the control for any nonlinear plant, even for the tracking control, and some limitations should be taken into account. To clarify these constraints, let us review the CD procedure starting from simple scalar models.

Assume a scalar plant with model

$$\dot{x}(t) = f(x(t)) + g(x(t))u(t) \tag{8.2}$$

and a reference signal $x_r(t)$, which is continuous and derivable, its derivative being also available. If the control goal is to generate a control signal to be applied to (8.2) to follow this reference, with a proportional approaching, the control law will be

$$u = \frac{\dot{x}_r + k(x_r - x) - f(x)}{g(x)} \tag{8.3}$$

as far as it is realizable. Clearly, this control law is composed of a feedforward term from the reference, and its derivative as well as a feedback term from the system state, $x(t)$. It can be easily proved that the tracking error tends to zero and the controlled plant is stable. In fact, by applying the control (8.3) to the system (8.2), it leads to

$$\dot{x} = \dot{x}_r + k(x_r - x) \tag{8.4}$$

and the error ($e_x = x_r - x$) tends to zero as far as $k > 0$. Now, a second-order model with a single input is assumed. Thus, only the reference for one state variable could be defined. Let us denote by $x_r(t)$ this variable, the second one being denoted as $z(t)$. The state space model can be expressed as

$$\begin{aligned} \dot{x} &= f_1(x,z) + g_1(x,z)u(t) \\ \dot{z} &= f_2(x,z) + g_2(x,z)u(t) \end{aligned} \tag{8.5}$$

If the control law is derived like in (8.3) from the first state equation, the behavior of the second state variable is unpredictable

$$\dot{z} = f_2(x,z) + g_2(x,z)\frac{\dot{x}_r + k(x_r - x) - f_1(x,z)}{g_1(x,z)} \tag{8.6}$$

and it could be unstable. In order to guarantee a stable behavior of the second state variable, a fictitious reference to be tracked should be generated, being compatible with the model behavior (8.5). Thus, the controlled plant model will be

$$\begin{bmatrix} g_1(x,z) \\ g_2(x,z) \end{bmatrix} u = \begin{bmatrix} \dot{x}_r + k_1(x_r - x) - f_1(x,z) \\ \dot{z}_r + k_2(z_r - z) - f_2(x,z) \end{bmatrix} \quad \Rightarrow \quad Au = b \tag{8.7}$$

This implies to choose the second reference in such a way that the vectors A and b become parallel.

$$\frac{g_1(x,z_r)}{g_2(x,z_r)} = \frac{\dot{x}_r + k_1(x_r - x) - f_1(x,z_r)}{\dot{z}_r + f_2(x,z_r)} \tag{8.8}$$

To obtain the explicit value of this reference is very complicated in the general case, but it could be simple in some (frequent) cases.

For instance, if $g_1(x,z) = 0$, the reference can be computed as the solution of

$$f_1(x,z_r) = \dot{x}_r + k_1(x_r - x) \tag{8.9}$$

and the control law would be

$$u = \frac{\dot{z}_r + k_2(z_r - z) - f_2(x,z_r)}{g_2(x,z_r)} \tag{8.10}$$

If there are the same number of control inputs than variables to be tracked, the situation is similar to (8.1) and (8.2), and the control vector law would be

$$u = G(x)^{-1}[\dot{x}_r + k(x_r - x) - F(x)] \tag{8.11}$$

where G and F are matrices of appropriate dimension, and k is a diagonal matrix of controllers' coefficients.

If the number of inputs (m) is lower than the process dimension (r), only a maximum of m process variables can be tracked, and the rest (the so-called sacrificed variables) will be used to smooth the reference tracking.

Again, the solution of (8.7) is the crucial point to reach an easy-to-implement control law, and the examples developed in the previous chapters illustrate the variety of processes where LAB CD methodology can be applied.

As an example of complexity, consider the reactor described in Chap. 6, with two control flows (reactor flow q and refrigerator flow q_j) and two variables to track, the reactor concentration and the temperature (Pérez & Albertos, 2004). The process model is

$$\underbrace{\begin{bmatrix} \frac{T}{V}(C_{ao} - C_{a,n}) & 0 \\ \frac{T}{V}(T_i - T_n) & 0 \\ 0 & \frac{\left(T_{j,n} - T_{ji}\right)}{V_j} \end{bmatrix}}_{A} \begin{bmatrix} q_n \\ q_{j,n} \end{bmatrix}$$

$$= \underbrace{\begin{bmatrix} C_{a,\mathrm{ref},n+1} - k_{C_a}(C_{a,\mathrm{ref},n} - C_{a,n}) - C_{a,n} + T\left(k_n C_{a,n}^2\right) \\ T_{\mathrm{ref},n+1} - k_T(T_{\mathrm{ref},n} - T_n) - T_n - T\left(-k_n C_{a,n}^2 \frac{H}{\rho c_p} - \frac{UA}{\rho c_p V}(T_n - T_{j,n})\right) \\ \frac{T_{j,\mathrm{ref},n+1} - k_{T_j}\left(T_{j,\mathrm{ref},n} - T_{j,n}\right) - T_{j,n}}{T} - \frac{UA}{\rho_j c_j V_j}(T_n - T_{j,n}) \end{bmatrix}}_{b} \tag{8.12}$$

From the first two rows, the value of $T_{j,\ \mathrm{ref},\ n}$ can be derived:

$$q_n = \frac{C_{a,\mathrm{ref},n+1} - k_{C_a}(C_{a,\mathrm{ref},n} - C_{a,n}) - C_{a,n} + T\left(k_n C_{a,n}^2\right)}{\frac{T}{V}(C_{ao} - C_{a,n})} =$$
$$= \frac{T_{\mathrm{ref},n+1} - k_T(T_{\mathrm{ref},n} - T_n) - T_n - T\left(-k_n C_{a,n}^2 \frac{H}{\rho c_p} - \frac{UA}{\rho c_p V}(T_n - T_{j,\mathrm{ref},n})\right)}{\frac{T}{V}(T_i - T_n)} \tag{8.13}$$

leading to

$$T_{j,\mathrm{ref},n} = \frac{T_i - T_n}{C_{ao} - C_{a,n}}$$

$$\frac{C_{a,\mathrm{ref},n+1} - k_{C_a}(C_{a,\mathrm{ref},n} - C_{a,n}) - C_{a,n} + T\left(k_n C_{a,n}^2\right) - \left(T_{\mathrm{ref},n+1} - k_T(T_{\mathrm{ref},n} - T_n) - T_n + T k_n C_{a,n}^2 \frac{H}{\rho c_p}\right)}{T \frac{UA}{\rho c_p V}} + T_n \tag{8.14}$$

As a result, the control flows are given by (8.13) and (8.15)

$$q_{j,n} = \frac{V_j}{T_{j,n} - T_{ji}} \times \left[\frac{T_{j,\text{ref},n+1} - k_{T_j}\left(T_{j,\text{ref},n} - T_{j,n}\right) - T_{j,n}}{T} - \frac{UA}{\rho_j c_j V_j}\left(T_n - T_{j,n}\right) \right] \quad (8.15)$$

Remark 8.1 The reference of the sacrificed variable (8.14), even complicated, is easy to implement. Also, its next value $T_{j,\text{ ref},\ n+1}$ required to compute the control action (8.14) can be estimated by extrapolating the previous sequence of values. But the references for the controlled variables, $C_{a,\text{ ref},\ n}$ and $T_{\text{ref},\ n}$ cannot be arbitrarily chosen because they can lead to unfeasible solutions, such as negative flows or out-of-range reference temperature for the refrigeration flow (8.14). Thus, the controlled variables' references should be admissible.

References

Besançon, G. (2007). Nonlinear observers and applications. In M. Thoma & M. Morari (Eds.), *Lecture notes in control and information sciences* (Vol. 363). New York: Springer.

Das, T., & Kar, I. N. (2006). Design and implementation of an adaptive fuzzy logic-based controller for wheeled mobile robots. *Control Systems Technology, IEEE Transactions on, 14*(3), 501–510.

Fernández, M. C., Nadia Pantano, M., Rossomando, F. G., Ortiz, O. A., & Scaglia, G. J. E. (2019). State estimation and trajectory tracking control for a nonlinear and multivariable bioethanol production system. *Brazilian Journal of Chemical Engineering, 36*(1), 421–437. https://doi.org/10.1590/0104-6632.20190361s20170379.

Li, T. H. S., Chang, S. J., & Tong, W. (2004). Fuzzy target tracking control of autonomous mobile robots by using infrared sensors. *Fuzzy Systems, IEEE Transactions on, 12*(4), 491–501.

Martins, F. N., Celeste, W. C., Carelli, R., Sarcinelli-Filho, M., & Bastos-Filho, T. F. (2008). An adaptive dynamic controller for autonomous mobile robot trajectory tracking. *Control Engineering Practice, 16*(11), 1354–1363.

Pantano, M. N., Serrano, M. E., Fernández, M. C., Rossomando, F. G., Ortiz, O. A., & Scaglia, G. J. (2017). Multivariable control for tracking optimal profiles in a nonlinear fed-batch bioprocess integrated with state estimation. *Industrial & Engineering Chemistry Research, 56*(20), 6043–6056.

Pérez M., & Albertos, P. (2004). Self-oscillating and chaotic behaviour of a PI-controlled CSTR with control valve saturation, *Journal of Process Control,* 14 5159.

Resende, C. Z., Carelli, R., & Sarcinelli-Filho, M. (2013). A nonlinear trajectory tracking controller for mobile robots with velocity limitation via fuzzy gains. *Control Engineering Practice, 21* (10), 1302–1309.

Rómoli, S., Amicarelli, A., Ortiz, O. A., Scaglia, G. J. E., & di Sciascio, F. (2016). Nonlinear control of the dissolved oxygen concentration integrated with a biomass estimator for production of *Bacillus thuringiensis* β-endotoxins. *Computers & Chemical Engineering, 93*, 13–24.

Rómoli, S., Serrano, M., Rossomando, F., Vega, J., Ortiz, O., & Scaglia, G. (2017). Neural network-based state estimation for a closed-loop control strategy applied to a fed-batch bioreactor. *Complexity, 2017*, 9391879. https://doi.org/10.1155/2017/9391879.

Sun, S., & Cui, P. (2004). Path tracking and a practical point stabilization of mobile robot. *Robotics and Computer-Integrated Manufacturing, 20*, 29–34.

Appendix A: Preliminary Concepts

Introduction

In this Appendix, the notation used in this book and the main basic concepts required to follow the developments are outlined. More details can be found in any basic linear algebra book, like Strang (1980) and Noble (1989), as well as in any basic control theory book, like Ogata (1995), Kalman (1960), Dorf and Bishop (2011), and Albertos and Sala (2004).

The content is organized as follows: first some mathematical preliminaries are drafted; then some basic results on dynamic systems theory are summarized.

Mathematical Preliminaries

The mathematical models used throughout the book to represent dynamic systems are mainly based on the internal representation. Thus, vectors and matrices will be used in an extensive way.

Vectors and Matrices

The representation of a set of variables as a vector has several advantages. Among them are the compactness and the geometrical interpretation of some vector operations. So, a vector can be represented as a geometrical object with a magnitude (module) and a direction. The number of components in the vector defines the dimension of the space where it can be drawn. For instance,

G. Scaglia et al., *Linear Algebra Based Controllers*,
https://doi.org/10.1007/978-3-030-42818-1

$$v = [v_1 \ v_2 \ \cdots \ v_r]^T \in R^r \tag{A.1}$$

represents a column vector (it could also be a raw vector if the transposition sign T is missing) where $R^r = V$ is the r-dimensional vector space, $\dim(V) = r$. Basic algebraic operations on real numbers, like addition, subtraction, negation, or multiplication by a real number, can be defined with vectors and they follow the same commutative, associative, and distributive laws as for real numbers.

Definition A.0 A vector is a *linear combination* of several vectors v_i if it can be expressed as

$$v = \sum_{i=1}^{n} a_i v_i \tag{A.2}$$

where a_i are real numbers.

Definition A.1 A vector is *linearly independent* (LI) of several vectors if it cannot be expressed as (A.2).

Definition A.2 A set of r linearly independent vectors like (A.1) define a vector space of dimension r.

Definition A.3 A set of $r_1 < r$ linearly independent vectors like (A.1) define a vector subspace of dimension r_1 in R^r.

As a conclusion, in a set of n r-dimensional vectors ($n > r$), there are at most r linearly independent.

Definition A.4 A *norm* $\|v\|$ of an r-dimensional vector like (A.1) is a real valued function with the properties:

$\|v\| \geq 0$ for all $v \in R^r$, and $\|v\| = 0$ if and only if $v = 0$;

$\|v + w\| \leq \|v\| + \|w\|$ for all $v, w \in R^r$,

$\|av\| = |a|\|v\|$ for all $v \in R^r$ and $a \in R$.

The most commonly used norm is denoted as *module or magnitude*, being

$$\|v\|_2 \triangleq \sqrt{\sum_{i=1}^{r} v_i^2} \tag{A.3}$$

Another usually used norm is denoted as *maximum*, being

$$\|v\|_\infty \triangleq \max_{i=1,2,\ldots r} |v_i| \tag{A.4}$$

Definition A.5 A *normed vector space* is a vector space where a norm has been defined.

Definition A.6 The distance between two vectors $u,\ v \in V$ is given by $d(u, v) = \|u - v\|$with the following properties:

1. $d(\mathbf{u}, \mathbf{v}) \geq 0$
2. $d(\mathbf{u}, \mathbf{v}) = 0$ if and only if $\mathbf{u} = \mathbf{v}$
3. $d(\mathbf{u}, \mathbf{v}) = d(\mathbf{v}, \mathbf{u})$
4. $d(\mathbf{u}, \mathbf{v}) \leq d(\mathbf{u}, \mathbf{w}) + d(\mathbf{w}, \mathbf{v})$

A sequence $\{u_i\}$ in a normed space V is denoted as a Cauchy sequence if $\|\mathbf{u}_n - \mathbf{u}_m\| \to 0$ when $n, m \to \infty$. A sequence converges to $\mathbf{u} \in V$ if $\|\mathbf{u}_n - \mathbf{u}\| \to 0$ when $n \to \infty$.

Definition A.7 An *inner product space* is a vector space with an additional structure called an inner product. This additional structure associates each pair of vectors in the space with a scalar quantity known as the inner product of the vectors. It is represented by $\langle .,. \rangle : V \times V \to F$ where V is a vector space, F is the field of real numbers, and $< \cdot\ , \cdot >$, the inner product, is a function satisfying the following:

1. $<\mathbf{u} + \mathbf{v}, \mathbf{w}> \ = \ < \mathbf{u}, \mathbf{w} > \ + \ < \mathbf{v}, \mathbf{w}>$
2. $<\alpha\, \mathbf{u}, \mathbf{v}> \ = \alpha < \mathbf{u}, \mathbf{v}>$
3. $<\mathbf{u}, \mathbf{u}> \ \geq 0, \quad < \mathbf{u}, \mathbf{u} > \ = 0$ if and only if $\mathbf{u} = \mathbf{0}$

Definition A.8 Two vectors are orthogonal if their inner product is zero. This property is represented as $u \perp v$.

Definition A.9 The *angle* θ between two vectors is defined as $\cos\theta = \frac{\langle u, v\rangle}{\|u\|\|v\|}$.

Definition A.10 Given a vector space V and a vector subspace U, $U \subset V$, for each $v \in V$ there exists $u \in U$ denoted as its *orthogonal projection* such that $\langle v - u, w\rangle = 0$, for all $w \in U$.

In a similar way, norms can be defined for functions.

Definition A.11 For a function of time, $x(t)$, $x : R \to R^r$, the *square norm* (or L_2 norm) is defined by

$$\|x\|_2 \triangleq \sqrt{\int_0^\infty \|x(t)\|^2 dt} \tag{A.5}$$

and the *infinite norm*, L_∞, as

$$\|x\|_\infty \triangleq \sup_{t>0} \|x(t)\| \tag{A.6}$$

Definition A.12 A function $f: R^n \to R^m$ is said to be *continuous at a point x* if given an $\varepsilon > 0$ a constant $\delta > 0$ exists such that

$$\|x - y\| < \delta \Rightarrow \| f(x) - f(y)\| < \varepsilon \qquad \forall x, y \in R^n$$

Definition A.13 A *matrix* is an array of vectors of the same dimension. The matrix dimensions are the number of vectors and their dimension.

For instance, an $m \times n$ matrix is an array of either n m-dimensional column vectors v_i or an array of m n-dimensional raw vectors w_i. A matrix can be represented as

$$A_{m\times n} = \begin{bmatrix} v_1 & v_2 & \cdots & v_n \end{bmatrix} = \begin{bmatrix} w_1 \\ w_2 \\ \vdots \\ w_m \end{bmatrix} = \begin{bmatrix} a_{11} & a_{12} & \dots & a_{1n} \\ a_{21} & a_{21} & \dots & a_{2n} \\ \vdots & \vdots & \vdots & \vdots \\ a_{m1} & a_{m2} & \dots & a_{mn} \end{bmatrix} \tag{A.7}$$

If $n = m$, the matrix is square.

Definition A.14 Between two vector spaces, V, W, there is *mapping* or transformation if there is a rule connecting each element of one vector space (denoted as *domain* of the mapping) with one element of the other (denoted as *codomain* of the mapping), its pair. If the mapping is invertible, for each element of the codomain there is a mapping (the inverse) assigning to it the paired element of the domain.

An $m \times n$ matrix A can represent a mapping of the vector space dimension n into the vector space dimension m, such that

$$W = A(V); \Rightarrow w = Av \tag{A.8}$$

According to the multiplication of matrices and vectors, for each element of w, it results in

$$w_i = \sum_{j=1}^{n} a_{ij} v_j \tag{A.9}$$

Definition A.15 A mapping is a *linear transformation* if the following properties hold:

$$\begin{aligned} A(v_1 + v_2) &= Av_1 + Av_2 = w_1 + w_2 \\ A(av) &= aA(v) = aw \end{aligned} \tag{A.10}$$

Definition A.16 The *rank* of a linear map between two vector spaces $A : V \rightarrow W$ is the order of the vector space W spanned by the map, also denoted as *image* of the map. For the A matrix, the rank is the maximum number of linearly independent column vectors.

Definition A.17 The *kernel (or null space)* of a linear map between two vector spaces $A : V \to W$ is the set of vectors such that $A(V) = 0$. For an A matrix, the dimension of the null space is the difference between the number of independent vectors and the total number of column vectors.

Thus,

$$\begin{aligned} \dim(\mathrm{Image}(V)) + \dim(\mathrm{null}(V)) &= \dim(V) \\ \mathrm{rank}(\mathrm{V}) + \mathrm{null}(V) &= \dim(V) \end{aligned} \tag{A.11}$$

For a square A matrix, $m = n$, the domain and codomain are the same, that is, $A : V \to V$.

Definition A.18 For a square matrix, an *eigenvector* v_e is defined as an element in V, different from zero, such that

$$Av_e = \lambda_e v_e \tag{A.12}$$

where λ_e is a real number denoted as its associated *eigenvalue*.

The set of eigenvectors define the image space of A. The eigenvalues can be obtained as the solution of (A.13):

$$|\lambda I - A| = 0 \tag{A.13}$$

which is denoted as the characteristic equation of A. I represents the identity matrix (a diagonal matrix with unitary elements).

Definition A.19 A square matrix is *invertible* or full rank (A^{-1}) if all its eigenvalues are different from zero and it is $A^{-1}A = I$. The n-column (raw) vectors are LI.

Definition A.20 A non-square matrix is full rank if the number of LI column (raw) vectors is equal to $\min\{n, m\}$.

Definition A.21 The *induced norm* of a matrix $A : V \to W$ is defined as

$$\|A\| = \sup\{\|Ax\| : x \in R^n; \quad \forall\|x\| = 1\} \tag{A.14}$$

Definition A.22 The *spectral norm* of a matrix A is the largest *singular value* of A, i.e., the square root of the largest eigenvalue of the matrix A^*A, where A^* denotes the conjugate transpose of A.

System of Linear Equations

Definition A.23 A *system of linear equations* (SLE) is a collection of two or more linear equations involving the same set of variables.

It can be represented by

$$\left.\begin{array}{l} a_{11}x_1 + a_{12}x_2 + \ldots + a_{1n}x_n = y_1 \\ a_{21}x_1 + a_{22}x_2 + \ldots + a_{2n}x_n = y_2 \\ \ldots \\ a_{m1}x_1 + a_{m2}x_2 + \ldots + a_{mn}x_n = y_m \end{array}\right\} \Rightarrow Ax = y \tag{A.15}$$

easily described in compact form by means of vectors and a matrix, as a matrix equation. The SLE in (A.15) includes m equations, n variables, and mxn coefficients, arranged as the elements of the A matrix. The y-vector is given (the data) and the x-vector is the SLE unknown variables (or solution).

Definition A.24 A *solution* of a SLE is a vector x whose elements simultaneously satisfy the m equations in (A.15).

A SLE may have one solution, infinite number of solutions, or no solution at all. If the number of equations (data) is lower than the number of unknowns, there may be infinite solutions. If the number of equations is larger than the number of unknowns, a solution is not guaranteed. If $m = n$, a single solution is foreseeable. Looking at the SLE in (A.15), the existence of solutions depends on the A matrix dimensions.

(a) If $m = n$, the A matrix is square. If it is full rank, from (A.15), the following can be obtained:

$$x = A^{-1}y \tag{A.16}$$

This implies that (1) the m equations are LI, that is, the m raw vectors $\begin{bmatrix} a_{i1} & a_{i2} & \cdots & a_{in} \end{bmatrix}$ are LI and (2) the n column vectors $\begin{bmatrix} a_{1i} & a_{2i} & \cdots & a_{ni} \end{bmatrix}^T$ are also LI.

(b) If $m > n$, there are more equations than unknown variables. Thus, it is not guaranteed that a solution exists. An approximated solution can be found, if A is full rank, as follows:

$$A^TAx = A^Ty \Rightarrow x = \left(A^TA\right)^{-1}A^Ty = A^{\dagger}y \tag{A.17}$$

because A^TA is invertible. $A^{\dagger}$ is denoted as the left pseudoinverse matrix.

It can be shown that this solution minimizes the module of the solution vector.

(c) If $m < n$, there may be infinite solutions. Again, if A is full rank a possible solution may be obtained as follows:

$$\left.\begin{array}{c} x = A^T v \\ Ax = y \end{array}\right\} \Rightarrow \left.\begin{array}{c} AA^T v = y \\ v = \left(AA^T\right)^{-1} y \end{array}\right\} x = A^T \left(AA^T\right)^{-1} y = A^\dagger y \tag{A.18}$$

In this case, $A^\dagger$ is denoted as the right pseudoinverse matrix and, as before, the solution vector has minimum module.

(d) If min(m,n) < rank(A) < max(m,n), that is, A is not full rank, a generic pseudoinverse can be computed by means of the singular value decomposition. Thus, it is

$$A = U \sum V^T \Rightarrow A^\dagger = V \sum^{-1} U^T \tag{A.19}$$

Least Square Solution

In solving the SLE, where the solution x is underdefined, let us denote as equation residual r_i the difference between the value of the i equation for the approximated solution and the assigned data y_i. In vector notation it would be

$$r = Ax - y \tag{A.20}$$

The *least square* (LS) solution of (A.15) is the solution minimizing the sum of the squares of the residuals, that is,

$$x_{\text{opt}} = \min_x \left(\sum_i r_i^2 \right) = \min_x \left(r^T r \right) \tag{A.21}$$

Taking into account that, in matrix notation, $\frac{d(y^T Ax)}{dx} = A^T y$, $\frac{d\left(x^T A^T y\right)}{dx} = A^T y$, and $\frac{d(y^T y)}{dx} = 0$, the minimum in (A.21), where the residuals are defined in (A.20), will be

$$\frac{d(r^T r)}{dx} = 2A^T Ax - 2A^T y = 0 \Rightarrow x_{\text{opt}} = \left(A^T A\right)^{-1} A^T y = A^\dagger y \tag{A.22}$$

where $A^\dagger$ is the left pseudoinverse matrix. This solution is denoted as the *ordinary least square* solution, and all the residuals receive the same relevance.

Weighted Least Square Solution

In some cases, not all the equations have the same confidence and a weight can be assigned to each residual. In this case, the optimal solution minimizes the weighted sum of the residuals, that is,

$$x_{\mathrm{opt}} = \min_x \left(\sum_i w_i^2 r_i^2 \right) = \min_x \left(r^T W r \right) \tag{A.23}$$

where W is a diagonal matrix with elements w_i^2. Following a similar reasoning as before, the optimal solution will be given by

$$\frac{d(r^T W r)}{dx} = 0 \Rightarrow x_{\mathrm{opt}} = \left(A^T W A \right)^{-1} A^T W y \tag{A.24}$$

Data Interpolation

In many applications, the value of a function is only known for some points, being required to estimate the value of the function for intermediate points. In the simplest case, two points (y_1, x_1) and (y_2, x_2) $x_1 < x_2$, are known and the value of the function $y = y(x)$ should be estimated for $x \in \{x_1, x_2\}$. Assuming the continuity and monotonicity of the function, it should be $y_1 < y(x) < y_2$. If a line connecting the original points is drawn in the x–y plane, the estimation

$$\bar{y} = y_1 + \frac{y_2 - y_1}{x_2 - x_1}(x - x_1) \tag{A.25}$$

is denoted as the *linear interpolation*, and the difference between the actual value of the function and the interpolated value is the interpolation error, $e_y = y(x) - \bar{y}$. The equation of the line connecting both points is

$$\bar{y} = \frac{y_2 - y_1}{x_2 - x_1} x + \left(y_1 - \frac{y_2 - y_1}{x_2 - x_1} x_1 \right) = \alpha_1 x + \alpha_0 \tag{A.26}$$

For values of $x \notin \{x_1, x_2\}$ (A.25) gives the *linear extrapolation*.

It can be shown that given a set of n points, a polynomial function connecting all the points can be drawn. This is denoted as **polynomial interpolation**. The polynomial coefficients can be computed, in a similar way to (A.26) such that data fulfill the n instances of the polynomial, i.e.,

$$y = \alpha_n x_n + \alpha_{n-1} x_{n-1} + \cdots + \alpha_1 x_1 + \alpha_0 \tag{A.27}$$

Function Approximation

Other than interpolation or extrapolation, sometimes the value of a function should be estimated in the proximity of a given point. That is, knowing the function value $f(x_0)$ how to estimate the value of the function $f(x)$ if $|x - x_0| < \varepsilon$, being ε sufficiently small. The Taylor series expansion is a useful tool for that provided that the function derivatives exist. A Taylor series is an infinite series of terms such as

$$f(x) = f(x_0) + \sum_{i=1}^{\infty} f^{(i}(x)\big|_{x=x_0} \frac{(x - x_0)^i}{i!} \tag{A.28}$$

where $f^{(i}(x)\big|_{x=x_0}$ is the i derivative of the function at the given point. This series converges if the distance ε is in the interval of convergence for this function, that is, if

$$\lim_{i \to \infty} f^{(i)}(x)\big|_{x=x_0} \frac{(x - x_0)^i}{i!} = 0 \tag{A.29}$$

In these circumstances, the function can be approximated by the first terms of the series:

$$\begin{aligned} f(x) &= f(x_0) + (x - x_0) \frac{d}{dx} f(x)\bigg|_{x=x_0} + R_1(x) \\ f(x) &= f(x_0) + (x - x_0) \frac{d}{dx} f(x)\bigg|_{x=x_0} + \frac{(x - x_0)^2}{2} \frac{d^2}{dx^2} f(x)\bigg|_{x=x_0} + R_2(x) \\ &\ldots \end{aligned} \tag{A.30}$$

where $R_1(x)$ is the residual of order i. Usually, the first-order approximation is good enough if the interval of convergence is sufficiently small.

An interesting property of this series is that the approximation of the series after the deletion of the residual can be cancelled if the last term of the truncated series is taken at an intermediate point. That is, for example for a first-order truncation:

$$\begin{aligned} f(x) &\simeq f(x_0) \\ f(x) &= f(x_0) + (x - x_0) \frac{d}{dx} f(x)\bigg|_{x=x_\varepsilon}; \quad x_\varepsilon \in (x, x_0) \end{aligned} \tag{A.31}$$

Trigonometric Function Approximation

This kind of approximation will be used in the following chapters to deal with trigonometric functions. In particular, let us compute the second-order approximation of the sine and cosine functions:

$$\begin{aligned} \sin x &\simeq \sin x_0 + (x - x_0)\cos x|_{x=x_0} - \frac{(x-x_0)^2}{2}\sin x\Big|_{x=x_0} \\ \cos x &\simeq \cos x_0 - (x - x_0)\sin x|_{x=x_0} - \frac{(x-x_0)^2}{2}\cos x\Big|_{x=x_0} \end{aligned} \tag{A.32}$$

Thus, for small $|x - x_0| < \varepsilon$ the first-order approximation will be good enough:

$$\begin{aligned} \sin x &\simeq \sin x_0 + \cos x_0(x - x_0) \\ \cos x &\simeq \cos x_0 - \sin x_0(x - x_0) \end{aligned} \tag{A.33}$$

According to (A.31), (A.33) can also be written as

$$\left.\begin{aligned} \sin x &= \sin x_0 + \cos x_0(x - x_\varepsilon) \\ \cos x &= \cos x_0 - \sin x_0(x - x_\varepsilon) \end{aligned}\right|; \quad x_\varepsilon \in (x, x_0) \tag{A.34}$$

or

$$\left.\begin{aligned} \sin x &= \sin x_0 + \cos \underbrace{(x_0 + \lambda(x - x_0))}_{x_\lambda}\underbrace{(x - x_0)}_{\Delta_x} \\ \cos x &= \cos x_0 - \sin \underbrace{(x_0 + \lambda(x - x_0))}_{x_\lambda}\underbrace{(x - x_0)}_{\Delta_x} \end{aligned}\right|; \quad 0 < \lambda < 1 \tag{A.35}$$

This is the so-called Lagrange form of the Taylor series residual (T. Apostol 1967).

Dynamic System Models

There are many definitions of dynamic systems. Some of them are very formal, from a mathematical point of view, and can be found in specialized textbooks. For the purpose of our study, a **dynamic system** is a set of interconnected elements whose time evolution depends on the initial situation and the external actions applied to it. In a dynamic system there are many variables, internal (depending on the system) and external (provided by the environment).

A mathematical model of a dynamic system is a mathematical expression, usually a differential or difference equation, a set of them or an operator, relating the internal and external variables in such a way that if the initial conditions (the value of the internal variables at a given initial time) are known and the time evolution of the external variables is also known, the evolution of the internal variables can be computed. The variables are only function of time so the model is based on ordinary differential equations. For special distributed systems partial differential equations should be used and the models are more difficult to handle.

The time evolution can be considered *continuous time*, when the time instant is a real number, or *discrete time* when the time is only considered at some prescribed time instants (the time is nT, where n is an integer, and T is the sampling period).

Among the external variables two different groups can be distinguished: the *control variables*, able to be manipulated to influence the system evolution, and the *disturbances* which are those external variables that in some cases can be measured but are determined outside the system.

A key concept in dynamic system modeling is the definition of *state*. The state of a system is the minimum number of internal variables allowing to compute all the system variables if the external variables are known. Those internal variables which are measured are denoted as *output* variables, providing information about the system situation. Usually, the external variables are denoted as *input* variables.

A system model is always an approximated representation of a system. An external model is a mathematical expression involving the input and output variables, whereas an internal model relates the input variables with the state and all them with the outputs.

A model is *linear* if the linearity properties hold: proportionality and superposition.

As seen in this chapter, a set of variables can be represented as a vector and thus the models may involve matrix equations.

External Models

An external model can be, for instance, a differential (difference) equation relating inputs and outputs and their derivatives (increments). These models, in the general case, are not very useful due to the difficulty to handle this kind of equations.

In the case of linear models, dealing with CT models, the Laplace transform allows to convert the differential equation into an operator, easy to handle. In this case the model is called the transfer function, and it can be handled as an operator with some physical properties. As an example, consider a dynamic system model given by the differential equation

$$\frac{d^n y(t)}{dt^n} + a_{n-1}\frac{d^{n-1}y(t)}{dt^{n-1}} + \cdots + a_1\frac{dy(t)}{dt} + a_0 y(t)$$
$$= b_m\frac{d^m u(t)}{dt^m} + \cdots + b_1\frac{du(t)}{dt} + b_0 u(t) \tag{A.36}$$

where $y(t)$ is the output and $u(t)$ is the input. If the Laplace transform is applied to (A.36), it yields

$$y(s) = \underbrace{\frac{b_m s^m + \cdots + b_1 s + b_0 u}{s^n + a_{n-1}s^{n-1} + \cdots + a_1 s + a_0}}_{G(s)} u(s) + Y_0(s) \tag{A.37}$$

where $G(s)$ is the transfer function and $Y_0(s)$ is the initial condition term. If this term is considered null, that is, the system is initially at equilibrium, the system model is just the transfer function.

For multi-input multi-output (MIMO) systems, inputs and outputs are expressed by vectors and the operator is a **transfer matrix**:

$$y(s) = G(s)u(s); \rightarrow G(s) = \begin{bmatrix} G_{11}(s) & \cdots & G_{1m}(s) \\ \vdots & \cdots & \vdots \\ G_{p1}(s) & \cdots & G_{pm}(s) \end{bmatrix} \tag{A.38}$$

A similar treatment can be done for DT systems by using the **Z** transform.

Internal Models

Internal models are composed of two equations: the state equation (A.39a) and the output equation (A.39b). For CT systems they are like

$$\begin{aligned} &\text{(a)} \quad \dot{x}(t) = F(x(t), u(t), t); \quad x(0) = x_0 \\ &\text{(b)} \quad y(t) = H(x(t), u(t), t) \end{aligned} \tag{A.39}$$

In (A.39) both equations are nonlinear and time varying. x_0 is the initial state and the input vector also includes the possible disturbances.

The model in (A.39) is very general but, again, difficult to handle. Usually, some assumptions are taken. For instance, for time invariant (TI) systems, the time is not an explicit argument.

For TI systems *affine* in the input, the state equation can be expressed as

$$\begin{aligned} &\text{(a)} \quad \dot{x}(t) = f(x(t)) + g(x(t))u(t); \quad x(0) = x_0 \\ &\text{(b)} \quad y(t) = h(x(t)) \end{aligned} \tag{A.40}$$

Finally, for *linear TI systems*, the model is reduced to

$$\begin{aligned} &\text{(a)} \quad \dot{x}(t) = Ax(t) + Bu(t); \quad x(0) = x_0 \\ &\text{(b)} \quad y(t) = Cx(t) + Du(t) \end{aligned} \tag{A.41}$$

where A, B, C, and D are matrices of appropriate dimension with real numbers.

Obviously, (A.41) represents the same system than (A.38); thus, both models are equivalent and easy to convert from one each other.

Discretization

Even if the system is a CT system, in order to study it and/or design some control, it could be interesting to work in the DT framework, either to use some design tools specially adapted for DT systems or to implement the control in a DT system. Thus, there are two options: design the control in CT and then look for its equivalent in DT or discretize the system model and design the controller in the DT framework.

A DT model of a CT system is always an approximation, as the information in the intermediate points is lost. Thus, a trade-off in discretizing a CT model is the selection of the *sampling period*, the time interval between two consecutive samples. If the sampling period is too big some relevant information can be lost. If the sampling period is too small, then the change between two consecutive samples could be negligible and the resulting model will be inaccurate. The Shannon theorem provides some guidelines about the selection of the sampling period but it is always a matter of expertise to decide what the best for a given application is.

Assuming a sampling period T, the discretization can be done following some rules, trying to preserve some properties of the CT model.

Derivative Approximation

For example, for simplicity reasons, the derivative terms can be replaced by the ratio of increments. That is,

$$\frac{dx}{dt} \cong \frac{x((n+1)T) - x(nT)}{T} \tag{A.42}$$

So, the Laplace variable s is substituted by

$$s \cong \frac{z-1}{T} \tag{A.43}$$

where z is the Z transform variable. This is denoted as the *Euler approximation* or one step-ahead approximation. If one step-back approximation is used, then the derivative is approximated as

$$\frac{dx(t)}{dt} \cong \frac{x(nT) - x((n-1)T)}{T}; \qquad s \cong \frac{1-z^{-1}}{T} \tag{A.44}$$

A more refined approximation is the so-called trapezoidal. This is a second-order approximation of the derivative, being replaced by

$$\frac{dx(t)}{dt} = f(t); \tag{A.45}$$

Taking into account the relationship between the Laplace and the Z variables, $z = e^{sT}$ this relationship can be approximated by the *bilinear (or Tustin)* approximation:

$$z = e^{sT} \cong \frac{1 + sT/2}{1 - sT/2} \Rightarrow \qquad s \cong \frac{2}{T}\frac{z-1}{z+1} \tag{A.46}$$

In all these cases, for the discretization of a CT transfer function, the equivalent DT transfer function is obtained by replacing s by its equivalent, that is, by (A.43), (A.44) or (A.46). Obviously, by the inverse relationships, a CT transfer function can be obtained from its equivalent DT transfer function.

Properties Preservation

Another approach to get the DT transfer function equivalent to a CT one is by preserving some of its properties. For instance, given $G(s)$, the step response equivalent DT transfer function $G_1(z)$ is the model providing the same output at the sampling instant when a step is applied at the input. This is equivalent to the relationship between the output sequence at the sampling instants and that of the input if it is piecewise constant over the sampling period, that is, if a zero-order hold device is implemented at the input, and it is denoted as the *zero-order hold equivalent*.

In the same way, given $G(s)$, the ramp response equivalent DT transfer function $G_2(z)$ is the model providing the same output at the sampling instant when a ramp is applied at the input. This is equivalent to the relationship between the output sequence at the sampling instants and that of the input if it is piecewise linear over the sampling period, that is, if a first-order hold device is implemented at the input, and it is denoted as the **first-order hold equivalent**.

The bilinear transformation yields a DT model with the same frequency response properties in a range of frequencies, in such a way that there is a correspondence between a frequency in CT ω_c and a frequency in DT ω_d, such that

$$\omega_c = \frac{2}{T} \tan\left(\omega_d \frac{T}{2}\right) \tag{A.47}$$

Another option is to preserve the zero-pole position. For a CT transfer function in factorized form, the pole-zero matching DT equivalent is given by

$$G(s) = k \frac{\prod_i s - z_i}{\prod_j s - p_j} \quad \Rightarrow \quad G_4(z) = \overline{k} \frac{\prod_i z - \overline{z}_i}{\prod_j z - \overline{p}_j} \tag{A.48}$$

where $\overline{z}_i = e^{z_i T}; \quad \overline{p}_j = e^{p_j T}$.

All the previous DT equivalents of the same CT transfer function converge if the sampling period tends to be zero.

Numerical Integration

The definite integral of a function can be computed digitally with some approximation. Given that

$$I = \int_a^b f(x)dx \tag{A.49}$$

to evaluate the integral the interval (a, b) is divided into small intervals and the function is assumed to be constant inside each interval, as shown in (A.50) and (A.51) and Fig. A.1:

$$I = \int_a^b f(x)dx \approx f(a)(b-a) = \text{Area of the rectangle aA1A2b} \tag{A.50}$$

or

$$I = \int_a^b f(x)dx \approx f(b)(b-a) = \text{Area of the rectangle aB1B2b} \tag{A.51}$$

Another option is to consider the linear approximation of the function, that is,

$$I = \int_a^b f(x)dx \approx \int_a^b g(x)dx = \frac{(b-a)}{2}[f(a) + f(b)]$$
$$= \text{Area of the trapezium aA1B2b} \quad (A.52)$$

denoted as trapezoidal approximation.

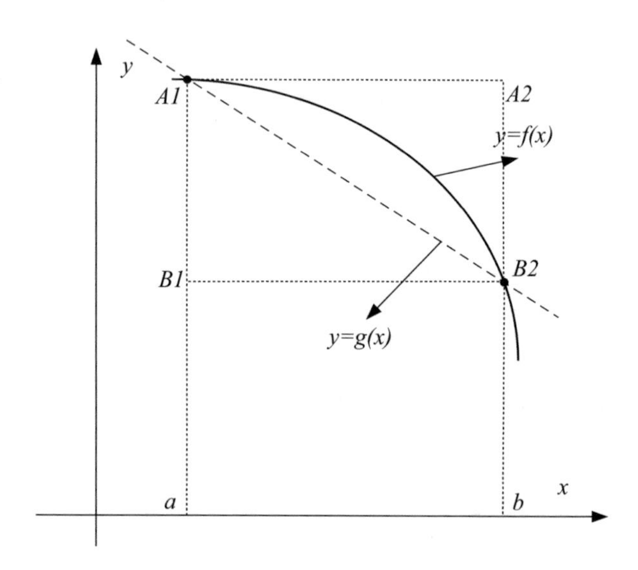

Fig. A.1 Numerical integration

References

Albertos, P., & Sala, A. (2004). *Multivariable control systems: an engineering approach*. Berlin: Springer Verlag.

Apostol, T. (1967). *CALCULUS, one-variable calculus, with an introduction to Linear Algebra* (p. 347). Waltham: Blaisdell Publishing Company.

Dorf, R. C., & Bishop, R. H. (2011). *Modern control systems*. London: Pearson.

Kalman, R. E. (1960). Contributions to the theory of optimal control. *Bol Soc Mat Mexicana, 5*(2), 102–119.

Noble, B. (1989). *Algebra Lineal aplicada*. Prentice Hall Hispanoamericana, Cop: México.

Ogata, K. (1995). *Discrete-time control systems* (Vol. 2, pp. 446–480). Englewood Cliffs, NJ: Prentice Hall.

Strang, G. (1980). *Linear algebra and its applications*. New York: Academic Press.

Index

G. Scaglia et al., *Linear Algebra Based Controllers*,
https://doi.org/10.1007/978-3-030-42818-1

Q

R

S

T

U

V

W

Printed in the United States
by Baker & Taylor Publisher Services